William Babcock Bellinger

A manual for Goldsmiths (1677)

William Babcock Bellinger

A manual for Goldsmiths (1677)

ISBN/EAN: 9783741104305

Manufactured in Europe, USA, Canada, Australia, Japa

Cover: Foto ©berggeist007 / pixelio.de

Manufactured and distributed by brebook publishing software
(www.brebook.com)

William Babcock Bellinger

A manual for Goldsmiths (1677)

A

TOUCH-STONE
FOR
Gold and Silver Wares.
OR,
A *Manual* for Goldsmiths,
AND
All other Persons, whether Buyers,
Sellers, or Wearers of any manner of
GOLDSMITHS Work.

DISCOVERING
The *Rules* belonging to that *Mystery*,
and the Way and Means how to know
Adulterated WARES from those made of the
True Standard Allay; And what are the True
Weights appointed for weighing of the same.

Together
With the several STATUTES now in
Force for Regulating Abuses committed in that
Craft. And the *CHARTER* of the Goldsmiths
Incorporation taken from the Record and truly ren-
dred into English.

To which is Annexed
The *LAWS* in force against Brass Hilts, and
Brass Buckles, &c. And Directions for Discovering the
Counterfeit Coyn of this *KINGDOM*. And also a
Catalogue of the Forraign Coyns, with the particular
Weights, Allay, and Value of each Coyn.

By *W.B.* of *London*, Goldsmith.

London, Printed for *John Bellinger* in *Cliffords-Inn* Lane,
And *Thomas Bassett* at the *George* near *Cliffords-*
Inne in *Fleet-street*, 1677.

The Intent of the Frontispiece.

1 *St.* Dunstan, *The Patron of the* Goldsmiths *Company.*
2 *The Refining Furnace.*
3 *The* Test *with Silver refining on it.*
4 *The Fineing Bellows.*
5 *The Man blowing or working them.*
6 *The* Test *Mould.*
7 *A Wind-hole to melt Silver in without Bellows.*
8 *A pair of Organ Bellows.*
9 *A Man melting or boyling, or nealing Silver at them.*
10 *A Block, with a large Anvil placed thereon.*
11 *Three Men Forging Plate.*
12 *The Fineing and other* Goldsmiths *Tools.*
13 *The* Assay *Furnice.*
14 *The* Assay-*Master making Assays.*
15 *His Man putting the Assays into the Fire.*
16 *The Warden marking the Plate on the Anvil.*
17 *His Officer holding the Plate for the Marks.*
18 *Three* Goldsmiths, *small-Workers, at work.*
19 *A* Goldsmiths *Shop furnished with Plate.*
20 *A* Goldsmith *weighing Plate.*

A 2   I Do

I Do, as far as in me lyes, Allow of the PRINTING of this BOOK.

Fra: North.

THE Author premiseth, That the matters comprized in the ensuing *Treatise* relating to Points of Law, or Expositions of any the *Statutes* therein mentioned, are not barely upon his own Opinion; But therein he hath taken the Judgment and Resolution of Councel Learned in the Law.

TO THE

RIGHT HONOURABLE,
Sir *Joseph Williamson* Knight and
Baronet, Principal Secretary of State
to *CHARLES* the Second, King of
Great *Brittain, France* & *Ireland*, &c.
and Lord Ambassadour and Plene-
potentiary for the Treaty of a General Peace at *Nimmegen.*

THe *subject of the ensu-
ing* Treatise *is* Gold
and Silver, *the Orna-
ment and* Riches *of this* Kingdome ; *And the design thereof is to make those* Metals *(in*

A 3 remo-

The Epiſtle

removing the Abuſes com-
mitted therein) to be really
ſo. Sir, although the Sphear
wherein I move in relation to
my knowledg of thoſe Metals,
hath rendred me capable to
make a real and plain demon-
ſtration of the Abuſes committed in the ſame, and what
remedies to apply ; yet without the favourable approbation
and aſſiſtance of thoſe in Eminent Places of Authority, it
may fail of the deſigned end,
which is to prevent the deceitful working of Gold and
Silver : The Conſideration of
which,

which, and for that I could
not apply my self to any person
for the Patronage of so Publique a Concern, more properly
than to your Honour, whose
Great Wisdom and Integrity
to the Publique-Weal, hath
advanced your Honour in His
Majesties Favour to Places
of greatest Eminency in the
State: Sir, knowing his Majesties Gracious Inclination,
and the intent of our Laws,
is for the securing his People
from injury; And that the
matter of the ensuing Discourse
may be no small Instrument

The Epistle

for effecting thereof in cases
relating to those Metals, I am
emboldened to make my humble Address to your Honour,
humbly imploring your Honours Favourable Acceptance
and Countenance of these my
weak Endeavours; And that
you will please to be Instrumental to enforce the due execution of the Laws in force
made for Regulation of the
working of Gold and Silver;
And where any debility shall
appear in those Laws, to supply the same by promoting some
more effectual Remedy for the
pre-

Dedicatory.

*preventing all Abuſes there-
in, and the advancing His
Majeſties Honour by the honeſt Manufacturing of Gold
and Silver, that therein we
be not inferiour to other Na-
tions, which is the hearty
Prayer, and ſhall be the earneſt
Endeavour of*

Right Honourable,

Your Honours humble

and devoted Servant,

W. B.

To the READER.

THere having not heretofore (that ever I could hear of) been expofed to Publique view any *Treatife* of this kind; This therefore may occafion more than ordinary Curiofity in Infpecting the Particulars thereof, and perhaps difpofe the *Reader* to the Confiderations following.

Firft, *Whether the Matter of it be true?*

Secondly, *What fhould incite me to this attempt?*

Thirdly, *Whether it be not againft the Rules of a Myftery thus publiquely to difcover them?*

Fourthly,

Fourthly, *Whether it may not (in-stead of the good propofed) become a difparagement to the Manufacture, and fo a prejudice to the Traders therein?*

Fifthly, *Whether this may not re-flect upon the Wardens of the Company, by fuppofing that they do not perform their Duty in fuppreffing thefe defects?*

As to the Firft.

I refer the *Reader* to the *Statutes* and other Laws (for Regulating the Goldsmiths Craft) herein cited, and his own obfervation, (from what is here intimated) how the fame have been purfued.

To the Second.

I have been moved hereunto from a defire of the Publique Good, Every honeft man's private benefit, and of detecting deceit and falfhood.

To the Third.

What I have here difcovered, is not the Honeft Myftery or Craft in Working and Fafhioning the Wares, (which in all Trades is to be conceal'd) but the Publique Rules of our Laws, and good Intentions of our Law-Makers to prevent Deceit in the Exercife of that Myftery which ill difpofed Craftsmen in all Ages have been too prone to contrive and practice.

To the Fourth.

It cannot be fuppofed, that for the offences of fome, all fhould be condemned ; Or that if fome Wares be deceitful, all fhould be fo made ; For I aver, That fome there be of this Trade (to their honour be it fpoken) that have not, nor will not in this Trade of a Goldfmith vary from the Rules and Principles of Law and Honefty, by working or felling adulterated Gold or Silver Works, though ftrongly befet with the temptations of
gains

gains for a more plentiful livelihood ;
and it cannot be denyed, that a perfon brought up in a Trade to live ho-
neftly thereon (to fee his Neighbours
draw away his Cuftomers by deceitful Wares under colour of Cheapnefs,
or lefler profit for his work, when indeed he makes greater advantages
thereby) and not be enfnared to do the
like practices, argues more than an ordinary integrity and conftancy.

To the Fifth.

It is well known the Wardens of *Gold-*
fmiths, London, have and will readily
hear all Complaints that fhall be made
to them therein by any perfon whatfoever, and punifh the Offenders : But
the Workers and Sellers in this Trade
being fo numerous, and the Tranf-
greffors fo fubtile to conceal their deceipt, and evade the punifhment, that
the endeavours of the Wardens, as yet
hath been, fo ftill will be but a weak
remedy, unlefs the buyer or Cuftomer
who wears and ufes the Commodity
be made capable of knowing the good
Wares

Wares from the bad, and the true value of either, and how to punish the offenders, which is the intent of the ensuing Discourse; Wherefore craving the favourable Construction of the Wardens and Company of the Mystery of Goldsmiths, *London* ; And all honest Workers and Sellers of *Goldsmiths* Work, for this my undertaking, I assure them and all my good Countrymen, That I value not the Calumnies of such who shall be displeased for discovering their deceitful practices; My only aym and intention in this Matter being to procure an honest Reformation in the making of Gold and Silver Works, and all other *Goldsmiths* Work whatsoever.

And having thus cast in my Mite in so just a matter, I have discharged my Conscience, and remain a devoted Servant to the Publique Good.

W. B.

The TABLE.

A.

(*)

B.

C.

                Cheap-

The Table.

D.

E.

F.

G.

Habita-

H.

I.

Keepers

K.

Marks

N.

O.

Officers

The Table.

Search

S.

(**)   Tables

T.

V.

The Table.

The Table.

A Touch-

A Touch-ſtone

FOR

Gold and Silver Wares:

OR,

A Manual for Goldſmiths.

That there hath been and is a great abuſe to the People of this Kingdome in the Silver and Gold that is wrought into the ſeveral ſorts of Wares in uſe amongſt us, is no hard matter to maniſeſt; For if there were inſpection

B   made

made into the Silver that is wrought
into Buckles for Belts, Girdles, Shooes,
Garters, and such like; And Hilts for
Swords and the pieces thereto belonging; And all other sorts of small work
both of Gold and Silver, and the value
thereof truly Examined, a great part
thereof would be found to want very
much of the value it ought to be of,
in respect both of the Standard, and
the Price it was sold at, whereby the
wearing buyer is abused and defrau-
ded, and the Lawes infringed, for the
private benefit only of deceitful and
ill-minded men. To direct therefore
in the Discovery of False Wares, and
to prevent the Evils arising thence,
I have framed this Discourse, which
consists of Three Parts; As,

Firft, *What Silver and Gold is in its
own Nature, and the Sort or
Standard, that is or ought to be
in use amongst us.*

Secondly, *A recital of the Statutes
and other Lawes in force for
preventing the working and
selling*

selling Gold and Silver Work,
that is worse than that Stan-
dard.

Thirdly, *Some* Notes *giving light to*
those Statutes, and directing
how to prevent Deceipt in Gold
or Silver work, and the way of
Redress for such Offences.

Silver is a Mineral of that Excellent
Nature, that when it is in the higheſt
degree of fineneſs, it will endure melting a long time in extream heat, with
but very little waſt, which quality
is not in any other Mettal, ſave Gold,
which (in the fineſt degree) is more
fixt, and will endure the fire with leſs
waſt: Wherefore Gold and Silver for
this excellency and their ſcarceneſs,
and capableneſs of being wrought into
ſo many ſorts of Uſeful and Ornamen-
tal things above other Mettals, is one
Reaſon that it bears ſo great a value,
and anſwers all things.

Our fore-fathers conſidering that
Silver in its fineſt degree would be too
ſoft

soft for ufe and fervice (for the fineft
Silver is almoft as foft as Lead), did
confult to reduce or harden the Silver
(by allaying it with bafer Metal) to
fuch a degree, that it might be both
ferviceable in the works, and alfo in
the wearing keep its native White-
nefs; And upon Experiment and due
Confideration, did agree that there
fhould be put Eighteen penny weight
of fine Copper into Eleven Ounces and
two penny weight Troy of the fineft
Silver, both which makes Twelve
Ounces or the pound Troy; And fo
according to that proportion for more
or lefs; (where it is to be obferved,
That either Tin, Pewter or Lead being
put into Gold or Silver for the allaying thereof, or being mixt therewith,
renders it extream brittle, and alto-
gether unfit for work) which degree
of allay is concluded upon by the Law-
makers of this Kingdome, to be the
Standard for all Silver Money, and all
Silver Works, and is commonly called the *Sterling Allay* (from the *Efter-
lings* or men that came from the *Eaft-*
Country, and were the firft Contrivers

vers and makers of that allay;) And this is that which is meant in the *Statute* of 18 *Eliz. Cap.* 15. by the Expression, (to wit) *Not less in fineness than that of* 11 *Ounces two penny weight*. And for this purpose divers *Statutes* have been made, which I have here recited *verbatim* as followeth.

The *Statute* 28 *Edw.* 1. *Cap.* 20.

It is Ordained, That no Goldsmith of England, nor none other-where within the Kings Dominions, shall from henceforth make or cause to be made any manner of Vessel, Jewel, or any other thing of Gold or Silver, except it be of good and true allay, (that is to say) Gold of a Certain touch, and Silver of the Sterling allay, or of better at the pleasure of him to whom the Work belongeth; and that none work worse Silver then Money; And that no manner of Vessel of Silver depart out of the hands of the Workers until it be Assayed by the Wardens of the Craft; and further, that it be marked with the Leopards-head:

B 3 And

And that they work no worse Gold then of the touch of Paris; And that the Wardens of the Craft shall go from Shop to Shop among the Goldsmiths to assay if their Gold be of the same touch that is spoken of before: And if they find any other then of the Touch aforesaid, the Gold shall be forfeit to the King; [And that none shall make Rings, Crosses, nor Locks,] And that none shall set any stone in Gold except it be Natural; And that Gravers or Cutters of Stones, and of Seals, shall give to each their weight of Silver and Gold as near as they can upon their Fidelity; And the Jewels of base Gold which they have in their hands, they shall utter as fast as they can; And from thenceforth if they buy any of the same Work, they shall buy it to work upon, and not to sell again: And that all the good Towns of England where any Goldsmith be dwelling, shall be Ordered according to this Statute, as they of London be; And that one shall come from every good Town for all the residue that be dwelling in the same, unto
London

These words of this *Statute,* [*None shall make Rings, Crosses, nor Locks,*]Repeal. 21 *Jac.* 28.

London, for to be ascertained of their
Touch. And if any Goldsmith be at-
tainted hereafter because that he hath
done otherwise then before is Ordained, he shall be punished by imprison-
ment and by ransome at the Kings
pleasure. And notwithstanding all
these things before mentioned, or any
point of them, Both the King and his
Council, and all they that were present
at the making of this Ordinance, will
and intend that the Right and Pre-
rogative of his Crown shall be saved to
him in all things.

Stat. 37 Ed. 3. Cap. 7.

Item, It is accorded, That Goldsmiths as well in London as elsewhere
within the Realm, shall make all
manner of Vessels and other Work
of Silver, well and Lawfully of the
allay of good Sterling; And every
Master Goldsmith shall have a mark
by himself, and the same mark shall be
known by them which shall be assigned
by the King to survey their Work and
Allay; And that the said Goldsmiths
B 4 set

ſet not their Marks upon their Works till the ſaid Surveyors have made their aſſay as ſhall be ordained by the King and his Councel; and after the Aſſay made, the Surveyor ſhall ſet the Kings Mark, and after the Gold-ſmith his mark for which he will an-ſwer; And that no Goldſmith take for Veſſel white and full for the weight of a pound, (that is to ſay) of the price of two Marks of Paris weight, but Eighteen pence as they do at Paris: [And that no Goldſmith making White Veſſel ſhall meddle with gilding,] nor they that do gild ſhall meddle to make white Veſſel: And they which ſhall be ſo aſſigned in every Town ſhall make their Searches as oftentimes as ſhall be Ordained; And for that which ſhall be in the Goldſmiths default, they ſhall incur the pain of forfeiture to the King, the value of the Metal which ſhall be found in default.

This Clauſe [*That no Goldſmith ſhall make white Veſſel and alſo gild*] Re-pealed 21 *Jac.* 18.

Stat.

Stat. 2 *Hen.* 6. 14.

The fineness of Harness of Silver, And that it shall be marked with the Leopard's *Head.*

Item, That no Goldsmith nor worker of Silver within the City of London, sell any workmanship of Silver, unless it be as fine as the sterling, except the same need Sodder in the making; which shall be allowed according as the Sodder is necessary to be wrought in the same; And that no Goldsmith nor Jeweller, nor any other that worketh Harness of Silver, shall set any of the same to sale within the City, before that it be Touched with the touch, and also with the Mark or Sign of the Workman of the same, upon pain of Forfeiture of the double value as afore is said: And that the Mark and Sign of every Goldsmith be known to the Wardens of the Craft. And if it may be found that the said Keeper of the Touch, touch any such Harness with the Leopard's
Head,

Head, except it be as fine in allay as
the sterling, that then the Keeper of
the Touch for every thing so proved
not as good in allay as the said sterling,
shall forfeit the double value to the
King, and to the party as is above
recited. And also it is likewise Ordained in the City of York, New-
Castle upon Tine, Lincoln, Norwich,
Bristow, Salisbury and Coventry,
That every one shall have divers
Touches according to the Ordinance
of the Mayors, Bayliffs, or Governours of the same Towns; And that
no Goldsmiths nor other Workers of
Silver, nor Keepers of the said Touches within the same Towns, shall set
to sale or touch any Silver in other
manner then is ordained, before within the City of London, upon pain of
the said forfeitures. And moreover,
That no Goldsmith or other Worker
of Silver within the Realm of England, where no touch is ordained as
afore is said, shall work any Silver except it be as fine in allay as the sterling; And that the Goldsmith or worker of the same Silver set upon the
same

same his Mark or Sign, before he set it to sale: And if it be found that it is not as fine as the sterling, that then the worker of the same shall forfeit the double value in manner and form as before is recited within the City of London. And the Justices of Peace, Mayors and Bayliffs, and all other having power as Justices of the Peace, shall here enquire and determine, by Bill, Plaint, or in other manner, all that do contrary to the said Ordinances, and thereof to make due execution by their discretions. Provided always, That if the Master of the Mint that now is, or which for the time shall be, offend or have offended in his Office of the said Mint, that then he be punished and Justified according to the form of the said Indentures.

Stat.

Stat. 18 Eliz. 15.

Whereas certain evil disposed Goldsmiths deceitfully do make and sell Plate and other Gold and Silver Wares to the great defrauding of her Majestie and her good Subjects; For Remedy whereof, Be it Enacted by the Authority of this present Parliament, That no Goldsmith from the Twentieth day of April next coming shall work, sell, Exchange, or cause to be wrought, sold, or exchanged, any Plate or other Goldsmiths Wares of Gold less in fineness then that of Twenty two Carrects, And that he use no Sodder, Ammel, or other stuffings whatsoever in any of their Works more than is necessary for the finishing of the same; And that they take not above the rate of Twelve pence for the Ounce of Gold (besides the fashion) more then the buyer may or shall be allowed for the same at the Queens Exchange or Mint, upon pain to forfeit the value of the thing so sold or exchanged: And that from the said Twentieth

tieth day of April, no Goldsmith shall
make, sell or exchange in any place
within this Realm, any Plate or Goldsmiths Wares of Silver, less in fineness then that of Eleven Ounces two
penny weight, nor take above the rate
of Twelve pence for every pound
weight of Plate or Wares of Silver,
(besides the fashion) more then the
buyer shall or may be allowed for the
same at the Queens Exchange or Mint,
Nor put to sale, exchange or sell any
Plate or Goldsmiths Work of Silver before he hath set his own Mark
to so much thereof as conveniently
may bear the same, upon pain to forfeit the value of the thing so sold or
exchanged: And if any Goldsmith shall
make any Goldsmiths Work or Plate,
and the same after the said Twentieth
day of April, shall be touched, marked
and allowed for good by the Wardens
or Masters of that Mystery, And if in
the same there shall be found any
falshood or deceit then the Wardens
and Corporation of that Mystery for
the time being, shall forfeit and pay
the value of the thing so exchanged
or

or sold. The one Moyety of all which Forfeitures shall be to the use of the Queens Majestie; and the other Moyety to the use of such party grieved and sustaining loss therby as will sue for the same in any Court of Record, by Action, Bill, Plaint, Information, or otherwise; wherein no Essoyne, Protection, or Wager of Law shall be admitted for the Defendant.

From which Statutes it is to be observed, That no manner of Silver Work whatsoever made, sold or exchanged in any place within this Realm, is to be worse than the aforesaid Standard or Sterling allay.

And for the better observing these Rules, the persons using that Mystery in and about the City of *London* have been and are Incorporated by the Name of the Wardens and Company of the Mystery of Goldsmiths of the City of *London*, and all that exercise that Mystery in the said City and Liberties thereof, ought to be of that Company; though many there are of that

that Myſtery who (through miſtake or deſign) are Free of other Companies, which yet hinders not but that they are to all intents and purpoſes in re-ſpect of their Works, as much under the power of the ſaid Company, as their own proper Members are; The Wardens thereof (I mean the Com-pany of Goldſmiths) are by the firſt recited Statute and their Charter, Au-thoriſed to ſearch amongſt all the Goldſmiths, and all others Trading in Gold and Silver Work in any place within this Realm, And to aſſay their Gold and Silver Work, and to break and deface all they do find of worſe allay than is appointed by the afore recited Statutes; And to fine the Of-fenders to the value of the Offence : Which large and copious Authority is known to have been put in practice : And for the Readers further know-ledge of all the Power and Authority to them given, I refer him to the In-rolment of the Patent now remaining of Record in the Chappel of the Rolls, a true Copy of which faithfully ren-dred into Engliſh, is hereunto annexed.

And

For the more easie discovering the
Workers and Sellers of unlawful Gold
or Silver Work, the Statutes aforesaid do appoint, That every Master
Worker in Goldsmiths Work within this Realm, shall have his proper
Mark, and the same Marks shall set on
their Works, before it be set to sale.

And that all such Workers Marks
(in the City of *London* and Three miles
compass of the same) to be known to
those assigned by the King to Survey
their Work and Allay, that is, the
Wardens of the Goldsmiths, And all
other Workers Marks in the several
places where Touches be ordained, to
be known to those appointed there to
Survey their work and allay, upon
the same penalty as is appointed for
working or selling course Silver-work,
(that is) to forfeit the value of the Silver work not marked with the Worker's Mark, or marked with a Mark not
so made known, *although the Silver be
of the fineness of Sterling.*

And by reason that under the general

ral term (*of any Goldsmith's Work*) mention'd in the Stat. of 18 *Eliz.* 15. is comprehended all Wier-work and Lace of Gold and Silver, which cannot be marked with the workers Marks (to answer the intent of the Law,) therefore our Law-Makers have put in this Proviso in the said Stat. 18 *Eliz.* 15. (*viz.*) to set the workers Mark to so much of his Work as will *conveniently* bear the same; but for all other Goldsmiths work, it will bear the Workers Mark with as much *conveniency* as with reason can be desired.

And for all Silver Work that is of the most Eminent account (of which are all sorts comprehended under the Names of *Vessels and Harness* that are made in and about the City of *London*, and within three miles of the same) these are not to be left unto or received upon the Credit and Reputation of the Maker thereof, by having only his Mark thereto; But the Credit and Reputation of the Company by their setting their Marks to the same, who are surely the most likely to continue,

C   and

and most able to make satisfaction in
case of defect, as is appointed by the
said Statutes, when the Worker and
Seller may be dead, or by several ways
disabled to make recompence to the
parties wronged.

The Company of Goldsmiths con-
sidering that their Wardens are by
their Charter and the Statutes aforesaid, appointed to Survey, Assay and
Mark the Silver-work, and that these
Officers are yearly chosen according
to their Usage out of their Members
of the Assistants, in course as they received their Admittance into those
Places; And that such Choice sometimes falls upon them that are either
of other Trades, or not Skill'd in that
Curious Art of making Assays of Gold
and Silver, and consequently unable
to make a true Report of the Goodness thereof, or else the necessary At-
tendance therein being too great a
Burden for the Wardens; Therefore
they have appointed an Assay-Master,
called by them their Deputy-Warden,
allowing him a considerable Yearly
Sallary,

Sallary, and who takes an Oath to this effect, (*viz.*) to perform that Office Faithfully according to the best of his Skill, that is, to make true assay of all Gold and Silver brought to their Office for that purpose, and elsewhere as the Wardens and Company shall appoint, and give a true Report of the goodness or badness of the same.

They have also caused to be made (according to the aforesaid Statutes and their Charter) Punchions of Steel, and marks at the end of them, both great and small of these several sorts following, (that is) *the Leopard's Head Crowned, the Lyon, and a Letter*, (a true Emblem of which Marks are expressed in the Copper Cut following) which Letter is changed Alphabetically every Year; the reason of changeing thereof is, (as I conceive) for that by the afore recited Statutes it is Provided, That if any Silver Work that is worse than Sterling be marked with the Companies Marks, the Wardens and Corporation for the time being shall make recompence to the party

C 2   grieved,

grieved; so that if any such default
should happen, they can tell by the
Letter on the Work in what Year it
was Assayed and marked, and thereby
know which of their own Officers
deceived them, and from them obtain
over, a recompence. These Marks are
every Year made New for the use of
the New Wardens; and although the
Assaying is referr'd to the Assay-Master, yet the Touch-Wardens looks to
the Striking the Marks.

They have also made in a part of
their Hall, a place called by them their
Assay-Office, as is before mentioned,
wherein is a Sworn Weigher, his Duty
is to weigh all Silver Work into the
Office, and enter the same into a Book
for that purpose; And also to weigh
it out again to the Owner, (only four
Grains out of every Twelve Ounces
that is marked, is according to their
Antient Custome to be detained and
kept for a re-assaying once in every
Year all the Silver Works they have
passed for good the year foregoing.

In

In this Office is likewise kept for
Publique View a Table or Tables artificially made in Columns (that is to
fay) one Column of hardened Lead,
another of Parchment or Velom, and
feveral of the fame forts ; In the Lead
Columns are ftruck or entred the
Workers Marks, (*which are generally
the two firft Letters of their Chriftian
and Sirnames*) and right againft them
in the Parchment or Velom Columns
are writ and entred the Owners
Names ; This is that which is meant
in the before recited Statutes, by the
Expreffion of *making the Workers Mark
known to the Surveyers, or Wardens of
the Craft :* Which faid Wardens Duty
is to fee that the Marks be plain, and
of a fit Size, and not one like another,
And to require the thus Entring the
faid Marks, And alfo the fetting them
clear and vifible on all Gold and Silver
work, not only on every Work, but
alfo on every part thereof that is
wrought apart, and afterwards Sod-
dered or made faft thereto in finifhing
the fame.

C 3   The

The reason for setting the Worker's Mark *on every part of the Work*, is to take away all colour of excuse from the maker of false work, who might otherwise craftily pretend that the part marked by him was good Silver or Gold, and (the unmarked part being bad) that the bad was added by some body else since the Plate went out of his hand.

Every Worker aforesaid is not only obliged to enter his Mark on the Table aforesaid, But according to the Companies By-Law for that purpose, is at the same time to enter in a Book (kept for that purpose in the *Assay-Office*) the place of his habitation; and if he remove to any other place, then to enter the same also in the Book aforesaid, so that their habitations may be always known to the Wardens of the Craft.

The Reason is, for that if at any time any Gold or Silver Wares be found to be of worse Allay then they should be, the Worker (by his Mark set thereon)

although

although he may not be known to the party grieved, yet by application to the Company, and by their comparing the Mark on the Work, to the Marks on their Table, may thereby be presently discovered.

Every Worker dwelling in the City of *London*, and Three miles compass of the same City, and also those in other places whose Marks are not so made known, according to the aforesaid *Statutes*, such Marks are deemed no Marks, and the making and selling Gold or Silver work whereon such fraudulent Marks are set, is as punishable as the working and selling Gold or Silver work that is worse than Standard as aforesaid.

And if the Companies Marks, or the Workers Marks that are made known to the Surveyers or Wardens as aforesaid, should be counterfeited on any Gold or Silver work, by any ways or means whatsoever, the Counterfeiter thereof is punishable for every such offence, by Indictment, and Fine to the

C 4   King,

King, and sometimes (as the Offence
may be circumstantiated) by the Pil-
lory; after which the Company or
Party whose Marks are Counterfeited,
may bring their Action against the
Offender, and recover the damage
sustained.

Our Law-Makers (as I conceive) did
think the thus setting the Marks on
the VVork, to be the securest way to
prevent Fraud in this kind; for if it
would not deter from the working
and selling Courfe Silver and Gold
Wares, yet would it be a sure way to
find out the Offenders, and to have
the injured righted: But if the Marks
might be omitted, and the work should
pass but into a third Owner's hands,
for the most part it would be impossi-
ble to difcern one man's VVork from
another, by reafon that divers workers
make all forts of VVork in shape so
neer alike.

It is queried by some, Whether the omitting to mark Silver Work that is Sterling, be punishable by Law, it being no positive deceit?

To which I anfwer.

That where a *Statute* commands a thing to be done, if the fame be omitted, it is a contempt of that Law, and punifhable; efpecially in this cafe where it may be confidered, That although it be not a direct deceit, yet to omit the marking good Silver work, is of ill example, and opens a door to deceit: for the permitting a worker to omit the fetting his Mark or the Companies Marks on his good work, is to encourage him at another time to work Courfe Gold or Silver, fince by fuch omiffion the Maker and Seller thereof will thereby remain undifcovered, and fo avoid making fatisfaction for his deceipt.

For the Difcovery of falfe Gold and Silver from that which is good, and to know the true value thereof, the manner

manner is this : The Assay-Master puts
a small quantity of the Silver upon
tryal in the fire, and then taking the
same out again, he with his exact
Scales that will turn with the weight
of the hundredth part of a Grain, com-
putes and reports the goodness or the
badness of the Gold or Silver.

In this Office are kept the Tools to
strike the Companies Marks, which
ought to be done clear and visible on
such Work that is Standard, and
what is worse ought to be broken and
defaced, whereby Thousands of Ounces of defective Silver yearly receives
execution; whence it may be inferred,
That if so many pieces of Plate and
Silver Work (that are brought to receive the Companies Marks) be broken and defaced, which were not suspected by the Owner, then surely the
Silver VVork never designed to be
brought to be Assayed and Marked,
much more deserves to be so served.

The said Company hath now appointed only Three dayes in every
Week

Week, (to wit) *Tuesdayes*, *Thursdayes*, and *Saturdayes*, (formerly every working day) to affay and mark the Silver Work ; And all perfons workers in Gold and Silver Works in the City of *London*, and Three miles of the fame City, are now (as by Law they ought to be) allowed to receive the Companies Marks on their Work ; and in cafe of the Workers neglect therein, the Wardens are to inforce the fame by going often on their Search, and break and deface all Gold and Silver Works finish'd and expofed to fale (among all Workers or Traders therein) that are not marked according to Law : Or at leaftwife where the Silver is Sterling, to Fine the Offender for fuch his unlawful neglect.

In the afore recited *Statute* of the 28 *Ed.* 1. *Cap.* 20. it is Enacted, That no manner of Veſſel of Silver depart out of the hands of the Workers, until it be Aſſayed by the Wardens of the Craft ; And further, That it be marked with the Leopards-Head. And in the afore recited Statute of the
2d.

2d. of *Henry* the 6th, Chap. 14. it is Enacted, That no Goldsmith nor Jeweller, nor any other that worketh harness of Silver, shall set any of the same to sale within the City of London, before that it be Touched with the Touch, and also with the Mark or Sign of the Workman of the same.

For the understanding of which, it is to be known, That all manner of Silver Work made to hold any liquid or other matter, is to be comprehended under and called by the general name of Vessels, although in the particulars, they are called by several other names ; As the Coyn of this Kingdom is called by the general term of Money; but in the particular parts thereof are called by several other names, as, Pence, Shillings, &c.

Under the title or term of *Harness*, (in the said Statute) is included all kind of Furniture for defence of Man and Horses against the Enemy, as Swords, Buckles for Belts, Girdles, and such like ; and also all other manner of of wearing Inftruments for Warr; which term of *Harnefs*, (in the fame fignification as we take it, by the Opinion of the beft Expofitors) is mentioned 2 *Chron.*9. 24. *Jer.* 46.4. *Exod.* 13. 18.

And that Hilts for Swords are comprehended under the term of *Harnefs*, as well as Buckles for Belts or Girdles, which may be collected by the *Statutes* 5 *Hen.* 4. 13. 3 *Ed.* 4. 4. 1 *Rich.* 3. 12. And 5 *Eliz.* 7. thofe Buckles are there called Harnefs, they being a material part of Girdles or Belts, and neceffary to compleat the fame for Martial Actions whereunto they are properly defigned. And a Girdle or Belt being of no ufe to that end without a Sword to be worn in the fame, and the Hilt being the chief defenfive part of that Weapon, the fame muft confequently be included under the term of Harnefs.

Which granted, it muft be concluded, That all Hilts of Silver, and Buckles of Silver for Girdles or Belts, &c. (being comprehended under the term

term of Harness) are by the recited
Statute of 2. *Hen. 6. Cap.* 14. to be
touched with the Touch, (that is) Affay'd and Marked by the Wardens of
the Goldsmiths.

Note, Their so doing was antiently
called the *Touch*; and the Wardens
that are to make the Assays and mark
the Silver, are now called the *Touch-
Wardens*.

By all which 'tis manifest, That all
Silver Hilts for Swords, and Silver
Buckles for Girdles or Belts, are not
only to be of the fineness of Sterling,
but also Assay'd and Marked by the
Wardens of the Goldsmiths before
they be exposed to sale, upon pain of
forfeiting double the value thereof.

From which I infer, (that the afore
recited *Statutes* positively appointing
all Vessels and Harness of Silver to be
marked with the Companies Marks)
The Wardens would have done well,
if (as touching Vessels and Harness)
they had omitted these Ambiguous
words in their late Precept (hereafter
inserted)

inferted) (*viz.*) if the said Works will conveniently bear the same; When in truth all the said Works can and must bear the same.

Therefore I conclude, That if the Wardens of the Goldsmiths shall be remiss in compelling all Workers of Vessels, and all Workers of Hilts for Swords, Buckles, and other Harness of Silver in the City of *London*, and three miles off the same, to bring the same Works to receive their Marks as aforesaid, it will be great Imprudence in them to lay themselves by such neglects open to the Law, when so small industry will prevent it.

In the Statute of the 2d. of *Henry* the 6th, *Chap.* 14. it is Enacted, That *Sodder* shall be allowed for the making up all Silver Wares (to wit) so much as is necessary for working the same. For the Explaining this word (Necessary,) see the Statute of 18 *Eliz.* 15. thereby the same quantity of Sodder is appointed again & further Enacted, That no Silver Work shall be worse in fineness nels then that of Eleven Ounces two penny weight; This is to be understood of all the parts thereof, besides the Sodder; for when all the Work is melted together, (that is) the foddered places with the rest, it will be worse then that Allay, by so much as the Sodder is worse than Standard; therefore the same Statutes though darkly, yet by a necessary Implication, limits the quantity of Sodder that shall be allowed for all Wares to a pennyworth in the Ounce, or a four penny weight in the Pound, by this Clause, **nor take above the rate of twelve pence for every pound weight** (that is, one pennyworth in every ounce) **of Plate or Wares of Silver besides the fashion, more then the buyer shall or may be allowed for the same at the Queens Exchange or Mint.** (That is)

If any Silver Work being melted, (with the Sodder thereof together) and it be more than the value of one pennyworth in the Ounce, or four penny weight in the Pound Troy worse than Sterling, it is not to be allowed.

And

And therefore if such have only the Makers mark thereon, the maker incurs the penalty of paying the value of the whole work ; and if the same be allowed and marked by the Company, they are to make satisfaction to the party grieved.

In the Statutes afore recited, the term, *Goldsmith,* is frequently used, for the understanding whereof it is to be known, that the working of Gold or Silver either by a private workman, or by the Master Shop-keeper , or his delivering out Gold or Silver to his Servants, or Workmen to be wrought into any sort of Work or Ware, either by *Melting, Filing or Hammering,* or every of them, is, or may (according to antient usage, and the intent of the afore recited Statutes) be deemed and taken to be the Trade of the *Goldsmiths.*

And every Person having served Seven Years Apprentiship, or as an Apprentice (to any Person that did before and during the said term follow the

D　　　　　　said

said Trade as aforesaid) is and may
properly be called a *Goldsmith*, and
such persons and their Apprentices
(and none other) may lawfully follow
or exercise the same; Stat. 5 *Eliz. chap.*
5.

Every worker of Gold or Silver
wares, as aforesaid , is by the intent of
the Law a *Goldsmith*, which appears by
the said Statute 18 *Eliz.* 15. which
principally aimed at the reforming the
abuse of making and selling deceitful
Gold and Silver work. For if the Makers of that Statute did not intend and
conclude all workers of Gold and Silver wares as aforesaid, to be *Goldsmiths*,
and all Workers and Retailers of deceitful Gold and Silver wares to be thereby punishable, that Statute (which is
the last made in that case) would be
ineffectual, and the abuse it intended to
reform , would notwithstanding re-
maine.

To conclude, if any Person hath bought
or received (of any worker or seller of
Silver work) any kind of Silver wares
suspected to be deceitful, the same
deceipt

deceipt may be known without doing prejudice to the work, by rubbing the Plate in some place least in sight, with a File of indifferent fineness, and if it be worse then Starling it will appear Yellowish, or else file it a little, and rub the Place filed on a cleane Touch-stone, and close by it rub the edge of a good Half-Crown-piece or such like thick money, and the difference, if any, will appear.

The reason that I direct the filing the Work is this (to wit) that the Artificial boiling of course Silver work, will so eat or dissolve the Allay that is on the surface or outside thereof, that unless it be filed as abovesaid, it will Touch on the Touch-stone six pence or eight pence in the ounce better then it is.

Note further, That to know a good Touch-stone, you must observe, That the best sort are very black, and of a fine grain, polished very smooth, and without any spungy or grain-holes; And near the hardness of a Flint, but yet with such a sharp cutting greet, that

Touch-stones are usually sold at the Iron-mong-ers in Fo-ster-Lane, London.

D 2

that it will cut or wear the Silver or
Gold when rubbed thereon.

The way to make a true Touch on
the Touch-stone, is thus; When your
Touch-stone is very clean, which if
foul or soily, it may be taken off, by
wetting it, and then rubbing it dry
with a clean Woollen Cloth ; or if
fill'd with Touches of Gold or Silver,
&c. it may be taken off by rubbing the
Touch-stone with a pumice-stone in
water, and it will make it very clean ;
then (your Silver being filed as abovesaid) rub it steadily and very hard on
the stone, not spreading your Touch
above a quarter of an inch long, and
no broader than the thickness of a
Five-shilling-piece of Silver ; And so
continue rubbing it until the place of
the stone whereon you rub, be like the
Metal it self: And when every sort is
rubbed on, that you intend at that
time, wet all the touch't places with
your Tongue, and it will shew it self in
its own countenance.

If

If it appear by thefe wayes to be worfe than Standerd, you may carry or fend it to the Goldfmiths *Affay-Office* aforefaid, and upon your defire, the Officers there will make an affay of the fame, and give you a true report of the value thereof in writing, and return the Ware(and Silver taken off for the Affay) to you again, no more defaced than what is done by the fcraping of the Silver for the affay.

But if you are minded to keep the matter more concealed, you may artificially cut or fcrape between 18 or 24 grains from fome one part, or from all the parts of the work (except the foddered places) (for lefs in weight than between 18 and 24 grains is not fufficient for an affay) Then in a piece of paper of about 6 inches long, and 4 inches broad, At the one end write down the Owners name, and the day of the Month and Year ; and at the other end put the cuttings or fcrapings of Silver in a fold, turning in the corners once, to prevent the fhedding the Silver, and fo fold up all the paper to

D 3.   the

the name so written, on the top as aforesaid. The manner and form for folding up the said Paper, and of the Assay-Master's Entry of his report in such Paper, is here presented by a Copper Cut.

Here place the Copper Cut.

Then carry or send it to the Goldsmiths Assay-Office as aforesaid (which is now on the South part of their Hall in *Foster-Lane, London*) on any of the Assay-days aforesaid, before the hour of 9 in the morning, and leave it with the Assay-Master or his Servant, and at 4 of the Clock in the Afternoon the same day it will be done; and by calling there for the Assay, by the name in the Paper, it will be delivered, upon the payment of 2 ^{d}, which is the accustomed Fee for the making of an Assay.

In which Paper the Assay Master will

John Dore's
' for an Assay Dec 18. 1675.

1

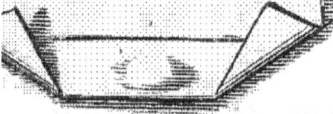

Will: Roe's
sil for an Assay Ian 2.st 1675.

Worse. iij. viij.

2

Tho: Dore's
ld for an Assay Apr. 10. 1675

Worse ij. iij.

3

Will Roe's
4

5   5   5

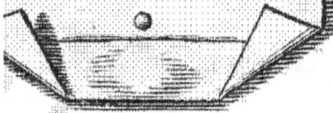

1 { An Assay paper open without
    the Assay Master's report.

2 { An Assay paper open of silver
    wth the Assay Masters report
    wch is i. 8. ob: in the ounce worse
    then standard silver.

3 { An Assay paper open of Gold
    wth the Assay Masters report
    wch is 10. 1.^d in the ounce worse
    then standard Gold.

4 { An Assay paper folded vp.

5 { The Company of Goldsmith's
    Marks. L being y letter for this
    year 1676.

will enter his report thereof in writing
in manner following,

If it be Sterling or Standard, ⎰ *Sta.*
  he will write —— —— ⎱

If it be a half-penny ⎰
  weight worse, he ⎱ *Worse ob.*
  will write——·——

                          dwt.
If it be a penny weight ⎰ *Worse* j
  worse, he will write ⎱

                    dwt.
If it be a penny weight ⎰
  and half worse, he ⎱ *Worse* j ob.
  will write, ——

                  dwt.
If it be Two penny ⎰
  weight worse, he ⎱ *Worse* ij
  will write, ——

And so proceeding higher,

                  dwt.
If it be Ten penny ⎰
  weight worse, he ⎱ *Worse* ꝟ
  will write, ——

If

If it be Ten penny ⎫   *dwt.*
   weight and a half ⎪ *worſe* ℧ *ob.*
   worſe, he will ⎬
   write, ———— ⎭

If it be Eleven pen- ⎫   *dwt.*
   ny weight worſe ⎬ *worſe* 𝔞𝔦
   he will write, —— ⎭

If it be Eleven ⎫   *dwt.*
   penny weight ⎪ *Worſe* 𝔞𝔦 *ob.*
   & a half worſe ⎬
   he will write, ⎭

And ſo proceeding higher.

If it be an Ounce ⎫
   worſe he will ⎬ *Worſe* j
   write, ——— ⎭

If it be an Ounce ⎫
   and a half penny ⎪ *Worſe* j *ob.*
   weight worſe, he ⎬
   will write, —— ⎭

If

If it be an Oun.
and 9 penny
weight worſe   *Worſe* j
he will write,

If it be an
Ounce &
19 penny
wt.worſe   *Worſe* j
he will
write,—— j

Note, That
[*dwt*] ſig-
nifies *penny
weight*,

and

ſignifies
Ounce.

And ſo proceeding higher according
as he finds the Silver courſer, and with
Numeral Letters ſetting down what
'tis worſe then Starling or Standard.

And it muſt be obſerved, whatever weight he ſets down, it is to be
accounted ſo much in every pound or
twelve Ounces Troy, and comes to this
effect (*viz*) that for every twenty
penny weight, or ounce Troy, that 'tis
reported worſe than Standard, you
muſt account ſix pence, and ſo propor-
tionable for more or leſs ; for ſo much
it will coſt for every ounce of ſuch
courſe Silver to make it of Starling
goodneſs, or to change it for Starling.

See more
Examples
of Allays
of Silver,
and the va-
lue thereof
caſt up, in
the Cata-
logue of
Forraign
Coyns, at
the latter
end inſert-
ed.

When

When you have so done, and your Silver being found and reported worse then Starling, you may make your complaint to the Master Warden of the Goldsmiths, he will cause the offender (living in or about *London*) to be Summoned to appear at the next Court of the said Company, and upon evidence of the Fact, the Wardens will (being obliged thereto by Law) procure you recompence, and punish the offender; they having promised so to do in their late Precept hereafter inserted.

If you dislike that way of proceeding, you may go by way of Action of Debt, Bill, Plaint, or Information, at the suit of the party grieved, who may sue in any of the Courts of Record at *Westminster*, and thereby recover the value of the whole deceitful or adulterated Wares, together with his charges, according to the said Statute of 18 *Eliz.*

There are also other ways of proceeding in these cases, as the afore mentioned Statutes do plainly direct.

                                        And

And that the perfons agrieved may be the better incouraged to proceed for their recompence, and to punifh the offenders, I have here inferted a true Coppy of an Indictment in Latin, and the fame rendred into Englifh, taken out of the Original (now remaining of Record in the *Crown-Office*) that was preferred (in *Trinity* Term, in the 28 year of the Reign of King *Charles* the Second *&c.*) before the Grand-Jury of Enqueft (attending the *Kings-Bench* Court at *Weftminfter*) againft a Perfon-offender in the premiffes ; which Bill (being drawn up, and this way of proceeding contrived by the advice of feveral Counfel learned in the Law) may ferve (with fome little variations, as occafion may be) for a good prefident in the like Cafes,

Per

Per Jud' Trin' xxviij.
Car' Secundi Regis.

Midd' ſſ. Jur̃ p Dño Rege ſuper Sacrm̃ ſuum pʒeſentant qd' A. B. nup de parocħ' Sc̃e M. in Com' pd' Aurifabac̃ qui educac̃ fuit in arte Aurifabʒoʒum ac ptres Annos & amplius jam ult' elapſ̃ eandem artem apud parocħ' pʒed' in Com' pʒedic̃' exercuit ac p totum idem tempus ib'm fuit liber homo Miſterij Aurifabʒoʒũ Civitat' London' ac quamplurima Uaſa & al' res ex argento confec̃' pʒetextu artis ſue pʒed ib'm fecit & fieri cauſavit ac diverſ̃ ligeis di̧ti Dñi Reg.s nunc ibidem Uendidit & Uenditioni expoſuit

Midd' ſſ. THe Jury for our Soveraign Lord the King upon their Oath do preſent, That *A.B.* late of the Pariſh of St. *Martin* in the Fields in the County aforeſaid, Goldſmith, who was brought up in the Trade of a Goldſmith, And for three years and more now laſt paſt hath exerciſed the ſame Trade within the Pariſh aforeſaid in the County aforeſaid, And for all that time there, was a Freeman of the Myſtery of Goldſmiths of the City of *London*, And there by colour of his ſaid Trade made and cauſed to be made

Ac

Ac qd' idem A.B. bene sciens qd' omnia Clasa & al' res ex argento confect & p homines Mistery Aurifabrorum Civitat London' venditioni expoit & exponend infra Civitat pd & alibi debent fieri undequacp [Anglicè wholly] de bono & vero Argento concordan cum Standard' Sancti dict Domini Regis et aut venditionem eorundem debeant assaiari [Anglicè be assayed] & signari [Anglicè marked] cum Insign [Anglicè the mark] Capitis Pardi [Anglice voc the Leopards Head] per Custod Misterij Aurifabrorū Civitat London' Quib⁹ Custod Insign ill' ad signand Clasa & al res de Auro & Argento

very many Vessels and other things of Silver, and there sold and exposed to sale to divers liegePeople of our said Soveraign Lord the King that now is; And that the said *A.B.* well knowing that all Vessels and other things made of Silver, and exposed and set to sale by men of the Mystery of Goldsmiths of the City of *London* within the City aforesaid and elsewhere, ought to be made wholly of good and true Silver agreeing with the Standard of the Exchequer of our said Soveraign Lord the King; And ought before the sale thereof to be assayed and marked by the Wardens of the Mystery of Goldsmiths

Con-

Concordat eam Standard prædict' confect' ante eorundem venditionem legittime creditur Idem tamen A. B. existens persona male & inhoneste dispositionis ac machinans nequit & fraudulent intendens ligeos & subdit' dicti Domini Regis nunc falso illicite deceptive callide & subdole decipere & defraudare & leges hujus Regni Angl' subvertere vicesimo primo die Januarij Anno Regni Domini nostri Caroli Secundi Dei Gratia Angl' Scot Franc & Hibern Regis Fidei Defensoris &c. Vicesimoseptimo apud Paroch' Scæ M. in Com præd' seraginf tibul' cingulat [Anglice coif vocat draw

of the City of London with the mark of the Leopards Head; which Wardens are lawfully entrusted with the Mark to mark Vessels and other things made of Gold and Silver agreeing with the Standard aforesaid before the sale thereof: Nevertheless the said A.B. being a person of an evil and dishonest disposition, and evilly devising and fraudulently intending the Liege People and Subjects of our said Soveraign Lord the King that now is, falsly, unlawfully, deceitfully, craftily and subtilly to deceive and defraud, and to subvert the Laws of this Kingdom of England, On the one and twentieth day of

Girdle

Girdle Buckles] & octo
fibul' Calcear' coif
Shooe Buckles de im-
puro & vilioii Argento
quam debet esse de ar-
gent' mie concordan
cum Standard poicto
videl't octodecim De-
nac in qualibet Uncia
inde Ulliorum quam
Argent' cum standard
prediceo concordan fal-
so fraudulent' & scienc
confecc' & fabricavit
Ac illa sic confecc' &
fabricac' ante aliquam
Assaraconem seu signa-
conem eorundem per
Cusood Misterii Auri-
fabrorum Civitac Lon-
don' pred' seu eorum
alterius postea scilicet
dco vicesimoprimo die
Januarii Anno regni
dicti Domini Regis
nunc vicesimoseptimo
supradicto apud Pa-
roch' sancte M.in Com

January, in the Se-
ven and twentieth
year of the Reign of
our Soveraign Lord
Charles the Second, by
the Grace of God
of England, Scotland,
France and Ireland
King, Defender of the
Faith, &c. at the Parish
of St. Martin in the
Fields in the County
aforesaid, falsely, frau-
dulently, and know-
ingly, did make and fa-
shion Threescore draw
Girdle Buckles, and
eight Shooe Buckles
of impure and baser
Silver than it ought
to be, not agreeing
with the Silver of the
Standard aforesaid,
that is to say, Eighteen
pence in every Ounce
thereof worser than
the Silver agreeing
with the Standard a-
pred

predict ut res ex Argento undequaqz conteck de bono & vero Argento concordan cu Standard predicto falso illicite deceptive & fraudulent venditioni exposuit ac divers ligeis diai Domini Regis nunc Jur predictis ignot adtunc & ibidem vendiconi exposuit & vendidit in contempt diai Domini Regis nunc legumqz suarum Ad grave dampnum & manifestam decepconem ligeorum predict qui hujusmodi fibulas cingular & fibul' Calceat emer In malum Exemplum omnium aliorum in consil' casu delinquen ac contra pacem dci Dni Regis nunc Coron & Dignitat' suas &c.

foresaid; And those things so made and fashioned before any assay or mark of the same by the Wardens of the Mystery of Goldsmiths of the City of *London* aforesaid, or either of them, Afterwards, that is to say, on the same One and twentieth day of *January*, in the aforesaid Seven and twentieth year of the Raign of our said Soveraign the King that now is, at the Parish of St. *Martin* in the Fields in the County aforesaid, as things made wholly of good and true Silver agreeing with the Standard aforesaid, falsely, unlawfully, deceitfully, and fraudulently exposed to sale; And then and there

there expofed to fale and fold to divers
Liege People of our faid Soveraign
Lord the King that now is, unknown
to the Jury aforefaid, In contempt of
our faid Soveraign Lord the King that
now is, and of his Laws, To the great
damage and manifeft deceit of the faid
Liege People who bought the faid
draw-Girdles-Buckles , and Shooe-
Buckles, to the evil Example of all
others offending in the like cafe,
Againft the Peace of our faid Soveraign Lord the King that now is, his
Crown and Dignity.

And fince thefe Directions are fo
plain, 'tis pity but he fhould be cheat-
ed that will not ufe them for preven-
tion, or to get recompence when deceived.

And my further Advice is, That
every perfon be careful to buy no Silver Work but what is marked as the
Laws require ; and if that proves
naught, recompence is eafily had; either of the Company (if marked with
their Marks, which (by reafon of the

E   care

care taken therein) are not ſet on Silver, worſe than Standard, in the compaſs of my Experience,) or the Maker
or Seller by his Mark will be eaſily
found out whereby to obtain recompence of him.

And in caſe of haſte, where the
Buyer cannot ſtay for the Companies
Marks, I adviſe him to take care that
he know the Workman to be able and
honeſt, and his Mark upon every part
of the Work that is wrought aſunder,
and afterwards ſoddered together as
aforeſaid.

*It is queried by ſome, Whether it be
Lawful for a Goldſmith to work
Gold or Silver, that is worſe than
Standard, if it be brought by the perſon, that is to receive it again, when
wrought up into Wares; or to work
it when 'tis beſpoke ſo to be, by the
perſon that will receive it when ſo
made up ?*

To

To which I anfwer by way of Cau-
tion and Advice,

That it is pofitively againſt the
the Laws afore recited to work Silver
or Gold, that is worſe than Standard,
into any ſort of Work under any pretence or colour whatſoever; and all
ſuch Wares be utterly unlawful, although the Worker do receive courſe
Silver to work for another, or receive
but the juſt value thereof; yet if it be
ſold or bartered to others, and happen
to be queſtioned, both the Maker and
Seller will be liable, and may be puniſhed as the Laws appoint.

The beſt Excuſe that can be made in
theſe caſes is, That ſuch courſe Work
is for the bringers or beſpeakers own
wearing: If that be true, the danger is
the leſs, becauſe the Worker lyes open
only to one; But his Wiſdome would
be the greater, not to lay himſelf open
to any.

Upon what is here declared, It is
E 2　　　　　hoped

hoped no person will suffer themselves to be deluded with the pretences of the Seller of unmarked and suspected Silver Work, (*viz.*)

That the Work will not bear the Marks as before is mentioned; for 'tis manifest and well known by great experience, That all Silver Works comprehended under the name of *Vessels*, and under the name of *Harness*, (Hilts for Swords and Buckles being included under that denomination) will bear the Marks appointed with as much conveniency as need to be desired, for the Vessels are generally marked with the bigger Marks aforesaid without exception; and for Silver Hilts and Buckles, (wherein the difficulty is supposed to lie) they have small Marks made on purpose for them, and the Work being first marked by the Worker with marks of Ink thus (o) on every place where the Workman thinks it most convenient to bear the Companies Marks; And the hollow Work being filled with Lead, (which is afterwards to be melted out again)

again) the Wardens will, or may, some
on the Anvil, and some on other Lead
which is put into a Vice for that purpose, strike their Marks on those places, both of wrought and plain Work,
without defacing or hurting the
same.

Nor need any person be deluded by
pretence that the Workmen have not
time to get it marked at Goldsmiths
Hall, when if the Work be carried to
the Office on any of the Assay dayes
aforesaid, before the hour of Nine in
the Morning, they may (if it be good
Silver) have it out ready marked at 4.
of the Clock in the Afternoon the
same day.

Or by pretence that the Work will
be so abused by the striking thereon
the Companies Marks, that it cannot
be finished Workman-like ; which indeed is one principal Excuse for not
bringing their Work to receive the
Companies Marks.

Neither let the supposition that the
Servants

Servants of the Company's Officers will abuse the Bringer or Owner of the Work either in word or demeanour (though some have heretofore been too rude and malipert) deter any from bringing their Works for the Company's Marks; for certain it is, the Wardens will not allow but severely reprehend their Officers and Servants that shall abuse any person or Work whatsoever.

And if the Wardens refuse to set their Marks, or not set them as they ought; or if they (or their Officers or Servants) shall do any damage to the Work by striking the marks, an Action well lyes against them; and they are besides by Law otherwise punishable and compellable to strike their Marks as Workman-like on the Work, as the Maker strikes his own mark thereon.

Nor let the Buyer suffer himself to be deluded at any time upon pretence that Standard Gold or Silver is too soft, and not so serviceable as that which is courser: For that pretence is vain,

vain, and 'tis well known by great experience, That both Gold and Silver of the Standard goodnefs well
wrought, into any forts of Wares, is
in every refpect better and more ferviceable than a courfer allay, which
moft times by reafon of the adultera-
tion, is found in the wearing not onely
of a braffy complexion, but very brittle and rotten.

Neither let the pretence of the Seller of adulterated Wares delude you,
(*viz.*) That he abates as much in the
fafhion as the Silver is allayd worfe
than Standard; when 'tis well known,
that moft times for every 6 *d.*he abates
in the fafhion of fuch courfe Silver
Work, he gains 1 *s.* 6 *d.* or 2 *s.* or more
by the allay in fuch Work.

Or by his promife, That he will at
any time allow five fhillings the Ounce
for the filver again, though it be broken to pieces: For by experience it is
evident, That very few Silver Wares
come again to be fold to the fame
hand, many being either kept to poft e-
rity;

rity, or transferred by gift; and if neceffity induce a fale, it is moft commonly in fome place remote from the place where it was bought, and then they muft take what they can get for it; which (if unmarked Wares) will not be more than 'tis worth.

I have made but little mention of GoldWares, and of the provifion made, to prevent deceit therein, becaufe Gold Wares are much lefs common than Silver Wares, I fhall therefore only propofe, That by the fame ways and method, by which you make difcovery of the goodnefs or badnefs of Silver, you may alfo make difcovery of the goodnefs or badnefs of Gold, and recover recompence if wronged, and punifh the Offenders; only with this difference, (viz.) That as the whiteft Silver is the beft, fo the Gold of the deepeft yellow is the beft; and the more the Gold inclining to a red or a pale yellow, 'tis fo much the courfer.

And as the Affay-Mafter in his reports

ports of the goodnefs of Silver fets it down by half-penny weights, and penny weights, and Ounces Troy, fo in his report of a Gold affay he fets it down by Carracts and Carract-grains, and half-grains.

For the underftanding of which you are to know, That five of the Troy grains makes a Carrect-grain, and four of fuch Carrect-grains makes one Carrect, and twenty and four of fuch Carrects, makes one Ounce Troy.

So that if Standard Gold be worth four pound the Ounce, for every carrect he fets down 'tis worfe, you muft account that in the Ounce Troy 'tis worfe by fo many times 3 *s.* 8 *d.* And for every grain he fets down 'tis worfe, you muft account it worfe by fo many times 11 *d.* in the Ounce Troy. And for every demy or half grain, 5 *d. ob.* for fo much it will coft to make it of Standard goodnefs, or to change it for Standard.

The manner of the Paper for a Gold affay,

assay, and the Assay-Masters report thereof is also expressed in the Copper Cutt aforesaid.

And further you are to know, That twenty and two of the aforesaid Carrects of the finest Gold, and two Carrects of fine Copper and Silver equal parts, makes an Ounce of Gold of the allay, appointed (by the Stat. of 18 of *Eliz.* 15. afore cited) to be the Standard for all Gold Wares, (worse than which allay no Wares are to be made, upon the penalty therein mentioned.)

And that 12 grains Troy is enough for making an assay of Gold.

But if any shall be dis-satisfyed with the assayings and reports of the Assay-Master of Goldsmiths Hall, may have assayes made by His Majestie's sworn Assay-Master of his Mint in the *Tower* of *London.*

Concerning Silver Work, made remote from *London,* I shall only insert, that

that it is to be obferved, That by the
firft recited Statute, all the Goldfmiths
in *England*, were appointed to bring
all their Silver Work (comprehended
under the name of Veffels) to *London*,
to be there affayed and marked with
the *Leopards* Head ; but the compelling thereof, under fuch great penalties as are therein mentioned, were
found to be a grievance : Therefore
by the Statute of 2 *Hen.* 6. 14. for the
better conveniency of the Goldfmiths
remote from *London*, Seven places are
appointed wherein fuch Work fhall
be affayed and marked, (*viz.*)

York, *Newcaftle upon Tine*, *Lincoln*,
Norwich, *Briftow*, *Salisbury*, and *Coventry*.

And as the Wardens of the Goldfmiths, *London*, are to affay and mark
the Silver Work that is made in and
about *London*, and three miles of the
fame, or to procure an Artift to do the
fame, (for which they muft anfwer)
fo, in every one of the aforefaid Seven
feveral places, the Chief Magiftrate or
Gover-

Governour is to aſſay and mark the
Silver Work that is made therein, or
to procure an Artiſt to do the ſame,
(for which they muſt alſo anſwer.)

, And in like manner as every Maſter-
worker in *Goldſmiths* Works in *London*,
and 3. miles compaſs of the ſame, are to
make their Marks known to the Wardens of the *Goldſmiths*, ſo every Ma-
ſter-Worker in Gold and Silver in every of the ſaid Seven Places, are to
make their Marks known to the Surveyors there (that is) to the Chief Ma-
giſtrate of ſuch Place : But what the
particular Marks that the reſpective
chief Governours of theſe Seven ſeveral Places ſet on the Silver work, I
can give no certain accompt thereof.

But this I can aſſert, That by reaſon the Marks of thoſe Places are little
known, they bear as little credit, and
therefore the *Goldſmiths* in remote places do frequently ſend up their Silver
work to receive the *London* Touch.

Here

Here followeth the Goldsmiths
CHARTER *truly rendred into*
English from the Copy thereof taken
from the Record now remaining in
the Chappel of the Rolls, *under the*
Title Confirmation, *Part the se-*
cond, Number the fourth, Confirmed
in the second Year of King James.

THE *KING* to all whom &c.
fendeth Greeting. We have perufed and feen the Letters Patents of
Confirmation of our Moft Dear Sifter
the Lady *Elizabeth* late Queen of
England, made in thefe words, *Eliza-*
beth by the Grace of God of *England,*
France and *Ireland,* Queen, Defender
of the Faith, *&c.* To all to whom thefe
prefent Letters Patents fhall come,
Greeting.

We have perufed the Letters Pa-
tents of Confirmation of the Lady
Mary late Queen of *England,* Our Moft
Dear

Dear Sister, made in these words,
Mary by the Grace of God of *England,*
France and *Ireland*, Queen, Defender
of the Faith, and on Earth over the
English and *Irish* Church the Supream
Head, To all to whom these presents
shall come, Greeting.

' We have perused the Letters Patents
of Confirmation of Our Dear Brother
Edward the Sixth, late King of *England,*
made in these words, *Edward* the Sixth
by the Grace of God of *England, France*
and *Ireland*, King, Defender of the
Faith, and on Earth over the *English*
and *Irish* Church, Supream Head, To
all to whom these present Letters shall
come, Greeting.

We have perused the Letters Pa-
tents of Confirmation of our Most
Dear Father, *Henry* the Eighth, late
King of *England* of Famous Memory,
made in these words, *Henry* by the
Grace of God King of *England* and
France, and Lord of *Ireland*, To all to
whom these present Letters shall come,
Greeting.

We

We have perufed the Letters Patents of Confirmation of our Moft Dear Father *Henry* the Seventh, Late King of *England* of Famous Memory, made in thefe words, *Henry* by the Grace of God King of *England* and *France* , and Lord of *Ireland,* To all to whom thefe prefent Letters fhall come, Greeting.

Know Ye, That We have perufed the Letters Patents of *Edward* the Fourth, Late King of *England,* made in thefe words, *Edward* by the Grace of God King of *England* and *France,* Lord of *Ireland,* To all to whom thefe prefent Letters fhall come, Greeting.

We have perufed the Letters Patents of *Edward* the Third, Late King of *England,* Our Progenitor, made in thefe words:

Edward by the Grace of God King of England, Lord of Ireland, and Duke of Aquitaine, To all to whom thefe prefent Letters fhall come, greeting.

Our Welbeloved the Goldfmiths of Our

Our City of London by their Petition exhibited to Us and Our Councel in Our Parliament holden at Westminster after the Feast of the Purification of Our Lady last past, have shewn,

That whereas no private Merchant nor Stranger heretofore were wont to bring into this Land any Money Coined, but Plate of Silver to exchange for Our Coyn.

And that it had been also Ordained, That all those who were of the Goldsmiths Trade were to sit in their shops in the High-Street of Cheap; and that no Silver in Plate, nor Vessel of Gold or Silver ought to be sold in the City of London, except at Our Exchange, or in Cheapside among the Goldsmiths, and that publickly, to the end the persons of the said Trade might Inform themselves whether the Seller came Lawfully by such Vessel or not.

But that now of late the said Merchants as well Private as Strangers, do bring from forraign Countries into this

this Nation Counterfeit Sterling,
whereof the pound is not worth above
sixteen Sols of the right sterling, and
of this Money none can know the true
Value, but by melting it down.

And also that many of the said Trade
of Goldsmiths keep Shops in obscure
turnings, and by-Lanes and Streets,
and do buy Vessels of Gold and Silver secretly, without enquiring if such
Vessel were stoln or lawfully come by,
and immediately melting it down, do
make it into Plate and sell it to Merchants Trading beyond Sea, that it
may be exported, and so they make
false Work of Gold and Silver, as
Bracelets, Lockets, Rings and other
Jewels; in which they set Glass of
divers Colours, Counterfeiting right
stones, and put more Allay in the silver
than they ought, which they sell to such
as have no skill in such things.

And that the Cutlers in their Work-
houses cover Tin with Silver so subtilly and with such sleight, that the same
cannot be discerned and severed from

F   the

the Tin, and by that means they sell the Tin so covered for fine Silver, to the great damage and deceipt of Us and Our People.

Whereupon the said Goldsmiths have Petitioned Us, That We would be pleased to apply convenient remedy therein.

And We being willing to prevent the said Evil, Do by and with the Assent of the Lords Spiritual and Temporal, and the Commons of Our Realm for the Common profit of Us and Our People, Will and grant for for Us, and Our Heirs,

That henceforth no Merchant either private or stranger, shall bring into this Land any sort of Money, But only Plate of fine Silver, nor that any Gold or Silver wrought by Goldsmiths, or any Plate of Silver be sold to the Merchant to sell again, and to be carried out of the Kingdom But shall be sold at Our said Exchange, or openly among the said Gold-smiths

smiths for private use onely.

And that none that pretend to be of the same Trade shall keep any Shop but in Cheapside, that it may be seen that their Work be good and right.

And that those of the said Trade may by vertue of these presents elect honest, lawful and sufficient men best skilled in the said Trade, to enquire of the matters aforesaid; and that they so chosen may upon due consideration of the said Craft reform what defects they shall find therein, and thereupon inflict due punishment upon the Offenders, and that by the help and assistance of the Mayor and Sheriffs if occasion be.

And that in all Trading Cities and Towns in England where Goldsmiths reside, the same Ordinance be observed as in London, and that one or two of every such City or Town for the rest of that Trade shall come to London to be ascertained of their Touch of Gold, and there to have a Stamp of a Pun-

cion

chion with a Leopard's Head marked
upon their Work as of antient time it
has been Ordained. In Witnesse
whereof We have caused these Our
Letters to be made Patents. Given
at Westminster the Thirtieth day of
March, in the First year of Our Reign.

We have also perused the Letters
Patents of *Richard* the Second after
the Conquest, late King of *England*,
made in these words, *Richard* by the
Grace of God King of *England* and
France, and Lord of *Ireland*, To all to
whom these presents shall come, greeting. Know Ye,

That whereas *Edward* our Grand-
father late King of *England*, at the
Suit of the Goldsmiths of our City of
London suggesting to him, how that
many persons of that Trade by Fire
and the smoke of Quicksilver, had lost
their sight, and that others of them
by their working in that Trade, be-
came so Crazed and Infirm, That they
were disabled to subsist, but by Relief
from others.

And that divers of the said City
Com-

Compaſſionating the Condition of such, were diſpoſed to give and grant divers Tenements and Rents in the ſaid City to the value of Twenty pounds *per Annum* to the Company of the ſaid Craft, towards the maintenance of the ſaid Blind, Weak, and Infirm; And alſo of a Chaplain to Celebrate Maſs amongſt them every day for the Souls of all the Faithful departed, according to the Ordinance in that behalf to be made, Did by his Letters Patents for the Conſideration of a Fine of Ten Marks, for himſelf and his Heirs, as much as in him lay, grant and give Licence to the Men of the Community aforeſaid, that they may purchaſe Tenements and Rents in the ſame City of the value of Twenty pounds *per Annum* and not above of the Men of that City, for relief and maintenance of ſuch blind and infirm, and of ſuch Chaplain as aforeſaid, to hold to them and their Succeſſors of the ſaid Society for ever, for the purpoſes aforeſaid, The Statute of *Mortmaine* or any other Statute or Ordinance to the contrary thereof notwithſtanding,

F 3

withſtanding, as in and by the ſaid
Letters Patents more fully and at large
it may appear.

And foraſmuch as the Men of the
ſaid Myſtery have humbly Petitioned
Us, That foraſmuch as Our Grand-
fathers ſaid Letters Patents are not
nor can be put in execution for want
of Naming Perſons capable therein,
That We would Gratiouſly provide
ſome remedy for them in this behalf.
We taking the Premiſſes into Conſideration, of Our eſpecial Grace, and for
the Conſideration of Twenty Marks
by them paid unto Us in Our Hana-
per,
Have for Us and Our Heirs granted
and given Licence to the men of the
ſaid Craft, That from henceforth they
be a perpetual Community or Society
of themſelves.

And that the ſaid Society or Company may for ever yearly Elect out of
themſelves four Wardens to overſee,
rule and duly govern the ſaid Craft,
and

and Community, and every Member
of the fame.

And further, We have according
to Our Authority in this behalf granted and given Licence for Us and Our
Heirs to the fame Wardens and Company, That they may purchafe and
have to them and their Succeffors,
Tenements and Rents, with their Ap-
purtenances, within the faid City and
Suburbs thereof to the value of Twenty pounds *per annum*, for the maintenance of the blind, weak and infirm
Men of the Company aforefaid, and of
a Chaplain to Celebrate Mafs amongft
the faid infirm, every day, for the Souls
of all the Faithful departed, for ever,
according to fuch Ordinance, As the
fame Wardens and Company fhall
make in this behalf, (the faid Statute,
or the Statute in that cafe made in
Our laft Parliament at *Weftminfter*,
notwithftanding,) or notwithftanding
that the faid Tenements and Rents be
held of Us in Free Burgage, fo that it
be found by Inquifition thereupon duly had and returned into our *Chancery*,

F 4   that

that such purchase may be made without any damage or prejudice to Us and Our Heirs, or any other person whatsoever. *In witness* whereof We have caused these our Letters to be made Patents, *Witness* Our Self at *Winchester* the Sixth day of *February*, in the Sixteenth Year of Our Reign.

And We ratifying and allowing the said Letters Patents, and all and every thing therein contained, do for Us and Our Heirs according to Our Power in that behalf Approve and Confirm the same, and do by these presents grant and Confirm the same unto Our Welbeloved the now Wardens and Company of the said Craft and their Successors for ever.

And of Our further Grace in this behalf, We for Us and Our Heirs, have granted to the same Wardens and Company, That notwithstanding they or their Predecessors have not hitherto upon any occasion in any sort used the Liberties in the said Letters Patents contained, Yet henceforth it
shall

ſhall be Lawful for them and their Succeſſors to Uſe and Enjoy the ſaid Liberties and every of them, without any Let or Impediment by or from Us or Our Heirs, or any of Our Juſtices, Eſcheators, Mayors, Sheriffs, Bay- liffs, or other Our Miniſters where- ſoever.

And We have further granted, and by theſe preſents do for Us and Our Heirs Grant to the ſaid now Wardens and Company of the Craft aforeſaid, That they and their Succeſſors be a Corporation or Body Incorporate, conſiſting and called by the Names of Wardens and Company, and be per- ſons able and capable in Law to pur- chaſe and take Lands and Tenements, Rents and other Poſſeſſions whatſo- ever, for ever in Fee-ſimple of any perſons whomſoever that ſhall be wil- ling to Give, Deviſe, or Aſſign the ſame to them,

To have and to hold the ſame to the ſaid Wardens and Company of the ſaid Craft, and their Succeſſors for ever.   And

And that they may and ſhall have
perpetual Succeſſion and a Common
Seal for the Affairs of their ſaid Myſtery.

And that they may by the name of
the Wardens and Company of the
Myſtery of Goldſmiths of the City of
London Implead and be Impleaded in
any Court and place whatſoever, before any Judges whomſoever, in any
Actions, Suits, Plaints, Demands and
Pleas as well real as perſonal or mixt
of what nature or kind ſoever they
be.

And that the ſaid Wardens and
Company and their Succeſſors, may
from time to time as often as they ſhall
judge it expedient, make good and
reaſonable By-Laws and Ordinances
for the better regulating the ſaid Myſtery,

We have alſo granted, and by theſe
preſents do grant to the ſaid now
Wardens and Company and their Succeſſors, That they may have and hold

to them and their Succeſſors for ever,
all and ſingular the Lands, Tenements,
Rents, Reverſions and Services heretofore given, granted or deviſed to
the Wardens of the Myſtery aforeſaid,
or to the Wardens or Guardians and
Company of the Myſtery aforeſaid, by
the Name of the Wardens and Company of the ſaid Myſtery, or by the
names of the Company of the ſaid
Myſtery, or by what other Name ſoever or heretofore in any manner pur-
chaſed by the Wardens and Company
of the ſaid Myſtery, or by the Company of the ſaid Myſtery, without Im-
peachment, Diſturbance or Let by or
from Us or Our Heirs, or by or from
Our Juſtices, Sheriffs, Eſcheators, or
other the Officers and Miniſters of Us
or Our Heirs whomſoever.

And We do hereby ratifie and con-
firm all and every the ſame Lands and
Tenements, Rents, Reverſions, and
Services to the ſaid now Wardens and
Company, and their Succeſſors.

And for the Credit of the Men of
the

the said Craft dwelling and residing in the said City for the time being, and for the preventing and avoiding the damage and loss which do or may daily happen and arise as well to Us as to any Our Liege People, for want of a due and provident Care in regulating certain of Our Subjects and others Using and Exercising the said Trade, *without any regard to the Credit* of the said Company, And also for the preventing and taking away Subtleties and *Deceipts practised in the said Trade,*

We have further granted, and by these Presents do grant to the said now Wardens and Company and their Successors for ever, That the Wardens of the said Mystery for the time being shall and may for ever have the *search, inspection, tryal and regulation* of all sorts of Gold and Silver wrought or to be wrought, and to be exposed to sale within the City of *London* and the Suburbs thereof, and in all Fairs and Markets, and all Cities, Towns and Boroughs, and all other places whatsoever *throughout Our Kingdom of England,*

*gland,*and also shall and may *have power to punish* and correct all defects that shall be found in the working of Gold and Silver.

And to that end, if need be, to call to their assistance the Mayor and Sheriffs of the said City, and the Mayors and Bayliffs or other Officers whatever in any Fairs, Markets, Cities, Boroughs and Towns, and other places out of the said City where any such Search or Tryal shall happen to be made.

And that the Wardens of the said Mystery for the time being shall and may have full Power and Authority for ever by themselves or any of them, duly to search and try all and singular the premisses, and also all manner of Work touching and concerning the said Craft, found or being in the hands of the Goldsmiths, *or any other whomsoever* selling, making or working any Wares or Works pertaining to or concerning the said Mystery, as well within Our said City of *London* and the

Suburbs

Suburbs thereof, as elsewhere out of the said City in all Fairs, Markets, Cities, Boroughs and Towns, and other places whatsoever throughout Our Realm of *England*; And also by themselves or any of them to *break* all such deceitful Works and Wares of Gold and Silver of what sort soever, if any such they shall find to be made, wrought and exposed to sale in deceipt of Our People.

And also according to their discretion and as often as they shall see it necessary to punish and correct the Makers, Sellers and Workers of the same Works according to their demerits, by the assistance (if need be) of Our Mayors, Sheriffs, Bayliffs, Reeves, and other such like Officers.

We also will and grant, and strictly Charge and Command, That all Bayliffs, Reeves, and other Officers whatsoever in Fairs, Markets, Cities, Boroughs, Towns and other places where such Search shall happen to be made, be ready to Ayd and Assist the said Wardens

Wardens and every of them making such Search as aforesaid, in the execution of the premisses, and that in all things according to their Duty. In witness whereof We have caused these Our Letters to be made Patents. Witness Our Self at *Leicester* the Thirtieth day of *May* in the Second Year of Our Reign.

And now We ratifying and approving the said Letters Patents, and all things therein contained, do for Us and Our Heirs according to Our Authority therein, allow the same, and by these presents do ratifie, grant and confirm the same to our Welbeloved the now Wardens and Company of the Mystery aforesaid and their Successors.

And further, We being certainly informed that divers persons both Natives and Aliens, exercising the said Trade in divers parts of this Kingdom, studying and contriving their own *dishonest Gain*, and purposing by various ways to *deceive* and endamage the

the reſt of our Subjects, do work and
expoſe to ſale Gold and Silver *cunning-*
ly and deceitfully wrought and debaſed
more than the Standard allowes, contrary to the Ordinances in that behalf
made.

And *Counterfeit-ſtones* (which are
of no value, cunningly ſet in ſuch kind
of Gold and Silver) do daily ſell for
true Jewels at great rates to divers of
Our Subjects (not underſtanding the
ſame) as well in places priviledged, as
in Fairs, Markets and other places
within Our Cities and Burroughs of
this Our Kingdom of *England,* and neither fear nor doubt to be puniſhed or
call'd in queſtion for the ſame.

And the reaſon is, for that due
ſearch or any due puniſhment is ſel-
dome executed upon Offenders in that
Myſtery out of the City aforeſaid.

And although the Wardens of that
Myſtery in the ſaid City for the time
being, have (by vertue and power of
the aforeſaid Letters Patents in form
aforе-

aforesaid granted to them and their Successors (had the search, inspection, tryal and regulation of the Gold and Silver so wrought and to be wrought and exposed to Sale, And power by themselves or any of them, to search and try (continually from the aforesaid 30th day of *May* in the said Second year of the late King *Edward* the Fourth hitherto by vertue of his said Letters Patents to them in that behalf made) all such kind of deceitful and fraudulent Works and Wares made and to be made of the Gold and Silver aforesaid of any kind whatsoever.

And the faults and deceipts in those Works deceitfully and subtilly contrived or to be contrived by the Workmen and contrivers thereof, to punish and correct, as also to execute and perform such other things as they ought to do by vertue of the aforesaid Letters Patents of the said late King *Edward* the Fourth;

Neverthelefs as We are informed, That notwithstanding the aforesaid

G   War-

Wardens of the said Mystery for the
time being have (ever since the aforesaid Grants to them and the Company
of the said Mystery in form aforesaid, made) been at great trouble and
charges (as well at their own proper
Costs, as at the Costs of the said Company) to put in execution their Authority of searching, inspecting and
trying such kind of Gold and Silver
(as is before mentioned) and the defects therein, so wrought and put to
sale by the Workmen of the said Trade
in divers of the Cities, Burroughs,
Fairs, Markets and other places of this
Our Kingdom, for the common use of
Us and of all Our good Subjects.

Yet they have received very little
or no profit thereby, but rather have
been subject oftentimes not onely to
pains and perils of their bodies, but
also to the loss of their Goods and
Chattels by reason and occasion of
their searching, trying, and putting in
execution their Authority aforesaid,
in punishing and correcting the defects
of Work upon proof thereof made
unto them,   By

By means whereof the said Wardens of late in regard of the great menaces and assaults which they have received from those Workmen and Tradesmen of that Mystery (that deceiptfully sell such Gold and Silver in the Countrey) and their Accomplices and Adherents, could not execute their said Authority any where within Our Kingdom of *England*, except within Our said City of *London* and the Suburbs thereof.

So that the said search, tryal and execution of the said Authority ceasing, very many frauds, deceipts, works unskilfully made of Gold and Silver, and of counterfeiting of Jewels in Works of Gold and Silver and otherwise, are daily divers wayes increased in the Mystery aforesaid, by the Workmen of the said Trade in every part of this Our Kingdom, to the great loss and detriment of Us and all Our Liege People,

And We being willing (all the deceiptful insufficient and unlawful

Works

Works and Wares of Gold and Silver
Jewels and Stones, Pearl or Coral, or
such like, in the Trade aforesaid un-
sufficiently and unlawfully made and
counterfeited used to be put to sale)
to abolish and punish in all things as it
ought to be, Of Our certain knowledg
and meer motion, Have given and
granted for Us and Our Heirs (as much
as in Us lyes) to the aforesaid Wardens
and Company of the Mystery aforesaid and their Successors,

That they the said Wardens and
their Successors, and every of them
for the time being, shall have for ever
full Power and Authority over all and
singular the Defects, Offences, Faults
and Deceipts made and attempted or
committed contrary to the Ordinances of the Mystery aforesaid in all their
searches and tryals of Gold and Silver,
or of Wares, Jewels, Stones, Pearl,
Coral, or of any other Jewels or Counterfeit Stones whatsoever wrought or
set in Gold or Silver, as in Neck-laces,
Lockets, Rings or Bracelets, or otherwise howsoever wrought or set, within

in this Our Kingdom of *England* wherefoever, againft the Workmen or Owners of any of the faid premiffes expofing them to fale.

And all and every the perfons of or ufing the faid Myftery whatfoever fo deceiptfully working, having, or expofing to fale the premiffes aforefaid, (upon due proof made) fhall be by the faid Wardens for the time being committed to the next Gaol or Prifon, therein to be punifhed by Imprifonment of their Bodies according to the nature of their Offences, and out of the faids Gaols or Prifons to be delivered at the difcretions of the faid Wardens or any of them; Or be punifhed by Fines to be fet and impofed upon fuch Delinquents, equal to their Offences, Faults and Deceipts, as by the found Difcretions of the faid Wardens or any of them for the time being, fhall be efteemed juft and reafonable, and in that behalf convenient.

We alfo grant for Us and Our Heirs to the faid Wardens and Company

and

and their Succeſſors, That whenſo-
ever, whereſoever, and as often as any
Wares of Gold and Silver or Pearl,
or of any Counterfeit Stones whatſoever deceitfully wrought or ſet in the
nature of Jewels or Pearl in Gold or Silver, which by Allays thereof are of leſs
value and more debas'd in the working of the ſaid Gold and Silver than of
right it ought to be wrought, (that is
to ſay) not being of the value of Sterling or Standard, (according to the
Ordinances and Statutes of Us and
Our Progenitors or Predeceſſors late
Kings and Queen of *England* in this
behalf made) that ſhall be found any
where as well within any of Our Liberties as without;

Or any Wares of Gold or Silver
made within this Kingdom by any Native or Forraign Workmen & Tradeſ-
men of the Craft aforeſaid whereſoever that ſhall be ſold or expoſed to
ſale (*not being tryed, approved and marked
as they ought to be*) according to the
form of the Ordinances and Statutes
aforeſaid, that then the ſaid Wardens
for

for the time being, or two of them,
shall have power and Authority all
and every such Wares of Gold and
Silver, Counterfeit Stones and Pearls,
and other Stones whatsoever (so deceiptfully or unlawfully wrought and
exposed to sale wheresoever they shall
be found) to arrest, seise, and to break
and spoyl them, so that Our People
may not be any more deceived thereby.

And that in all and every the
Searches of the said Wardens and
their Successors for the time being, of
or in the premisses from time to time
in whatsoever places within Three
Miles in and about the aforesaid City
of *London*, where any the said Workmen or Trades-men of the said Craft
shall happen to remain, work, or in-
habit, the said Wardens or any of them
for the time being shall cause to be
brought All manner of Works and
Wares of Gold and Silver aforesaid, or
what Jewels and Precious Stones soever (set in Gold and Silver) are there
wrought or to be wrought, to the

G 4 Com-

Common-Hall of the Wardens and
Company of the said Mystery being
in the aforesaid City of *London* wherein the Common Standard or Assise of
Gold and Silver (according to the Ordinances in that behalf made) is kept,
there to be tryed and assayed, And to
be reformed if defects shall be any
manner of wayes found therein; and
after they shall be so reformed, to be
there then affirmed for good, and
stamped with their Marks which they
use for that purpose.

And all defective Works whatsoever deceiptfully wrought as well of
or in Gold as Silver, Counterfeit
Stones put for Jewels therein and
falsly made, (or found to be of a worse
Allay than it ought to be) shall there
(according to their discretions) be utterly condemned; Without Accompt
or any other charge or Answer to Us
or any of Our Heirs for the premisses
or any of them to be rendred, made
or paid for ever; *In witness* whereof
We have caused these Our Letters to
be made Patents: *Witness* Our Self
at

at *Westminster* the Third day of *February*, in the Twentieth Year of Our Reign,

Now We Ratifying and Approving the said Letters Patents, and all things therein contained, Do for Us and Our Heirs (as much as in Us lyes) allow the the same, and by these presents do Ratifie, Grant and Confirm the same to Our Welbeloved the now Wardens and Company of the Mystery aforesaid. *In witness* whereof We have caused these our Letters to be made Patents, *Witness* Our Self at *Westminster* the Sixteenth day of *March*, in the First Year of Our Reign.

Now We Ratifying and Approving the said Letters Patents and all things therein contained, Do for Us and Our Heirs, as much as in Us lyes, allow the same, and by these presents do Ratifie, Grant and Confirm the same to Our Welbeloved the now Wardens and Company of the Mystery aforesaid ; *In witness* whereof We have caused these Our Letters to be made
Patents,

Patents, *Witness* Our Self at *Westminster* the Sixth day of *June* in the First Year of Our Reign.

Now We Ratifying and Approving the said Letters Patents and all things therein contained, Do for Us and Our Heirs as much as in Us lyes, allow the same, and by these presents do ratifie, grant and confirm the same to Our Welbeloved the now Wardens and Company of the Mystery aforesaid. *In witness* whereof We have caused these Our Letters to be made Patents, *Witness* Our Self at *Westminster* the Fifth day of *December*, in the First year of Our Reign.

Now We Ratifying and Approving the said Letters Patents and all things therein contained, Do for Us and Our Heirs, as much as in Us lyes, allow the same, and by these presents do ratifie, grant and confirm the same to Our Welbeloved the now Wardens and Company of the Mystery aforesaid ; *In witness* whereof We have caused these Our Letters to be made Patents. *Wit-*

Witness Our Self at *Westminster* the Third day of *January*, in the Third Year of Our Reign.

Now We Ratifying and Approving the said Letters Patents, and all things therein contained, Do for Us and Our Heirs, as much as in Us lyes, allow the same, and by these presents do Ratifie, Grant and Confirm the same to Our Welbeloved the now Wardens and Company of the Mystery aforesaid. *In witness* whereof, &c. *Witness* the King at *Westminster* the Thirtieth day of *March*, in the Second Year of the Reign of King *James* over *England*, &c.

THE

The Goldsmiths ORDER lately made and set forth for Prevention and Redress of the great Abuses committed in the several Wares afore-mentioned.

Goldsmiths-Hall the 23. *day* of *February,* 1675.

WHEREAS Complaint hath been made to the Wardens of the Company of Goldsmiths, London, That divers small Works, as Buckles for Belts, Silver Hilts, and the pieces thereto belonging, with divers other small Wares both of Gold

Gold and Silver, are frequently wrought and put to sale by divers Goldsmiths and others, worse than Standard, to the abuse of his Majesties good Subjects, and great discredit of that Manufacture; And that there are also divers pieces of Silver Plate sold, not being assayed at Goldsmiths-Hall, and so not marked with the Leopard's Head Crowned, as by Law the same ought to be: And whereas the Wardens of the said Company to prevent the said frauds, have formerly required all persons to forbear putting to sale any adulterate Wares either of Gold or Silver, but that they cause the same forthwith to be defaced; And that as well Plate-workers as small-workers should cause their respective Marks to be brought to Goldsmiths-Hall, & there strike the same in a Table kept in the Assay-Office; And likewise enter their Names and places of Habitations in a Book there kept for that purpose, whereby the persons and their marks might be known unto the Wardens of the said Company, which having not hitherto been duly observed,

observed, These are therefore to give
Notice to, and to require again all
those who exercise the said Art or Mystery of Goldsmiths in or about the
Cities of London and Westminster,
and the Suburbs of the same, That
they forthwith repair to Goldsmiths-
Hall, and there strike their Marks in
a Table appointed for that purpose,
and likewise enter their Names, with
the places of their respective dwell-
ings, in a Book remaining in the Assay-Office there: And that as well the
Worker as Shop-keeper, and all others
working and Trading in Gold or Silver Wares, of what kind or quality
soever they be, forbear putting to sale
any of the said Works, not being a-
greeable to Standard, that is to say,
Gold not less in fineness then two and
twenty Carracts, And Silver not less
in fineness then eleven Ounces two
penny weight; And that no person or
persons do from henceforth put to sale
any of the said Wares either small or
great, before the Workmans Mark be
struck thereon, And the same Assayed
at Goldsmiths-Hall, and there approved

ved for Standard, by ſtriking thereon
the Lyon and Leopard's Head Crowned, or one of them, if the ſaid Works
will conveniently bear the ſame: And
hereof all perſons concerned are deſi-
red to take notice, and demean themſelves accordingly: otherwiſe the Wardens will make it their Care to procure them to be proceeded againſt according to Law.

Touching the ſeveral Weights now
in uſe, for the buying and ſelling of
Gold and Silver and pretious Stones,
The *Reader* may Obſerve,

That no other Weights are (by our
Lawes) to be uſed in weighing Gold
and Silver, but thoſe called by the
name of Troy Weights, of which

24 grains makes a penny weight, or
the weight of an old Sterling penny, (which now goes for three
pence.)

20 penny weight makes one ounce.
12 Ounces make a pound.

The

The compounding these Weights (being used in Assaying of Gold, and computing the Standard of Gold) are called by the name *Carracts*, of which

5 of the aforesaid grains makes a Carract-grain; a demy-grain is half of such a grain.

4 of such Carract-grains, make one Carract.

24 of such Carracts make an Ounce Troy.

There be othor sorts of Carracts compounded of Troy grains, thus;

4 grains makes a Carract.

6 of such Carracts makes a penny weight.

120 of such Carracts makes an Ounce Troy; These are only used to weigh Diamonds and Pearls.

That all persons may know the difference

ference of Troy Weights from others,
they are to obferve that thefe Weights
are made in the fhapes, and of the Metal, and marked as hereafter is mentioned; (viz.)

The Grain Weights are made of pie-
ees of thin Brafs, commonly called Lat-
tin-Brafs, and are cut, near 4. fquare, and
proportioned from half a grain to fix
grains; and fo many grains that each
piece contains, it is marked with the
like number of round Marks thus (o)
And alfo on every piece is (or fhould
be) fet the letter G with a Coronet at
the head of it, thus (Ğ)

The next Weights above them, are
the penny Weights, which are made
of thick fquare pieces of Brafs, proportioned from a half-penny weight,
to a five penny weight, and not ufu-
ally higher; And fo many penny-
weight that each piece contains is
made or marked thereon, fo many
round marks thus (o) as abovefaid;
and alfo is or fhould be fet the *Lyon* on
every piece.

H   The

The next above them is the Ounce-Weights, they are also of Brass, and made round in nests, (that is to say) to fall or stand one within another, And are proportioned from a drachm to 32 Ounces, and sometimes to 64 Ounces, (*viz.*) the least is a †drachm; the next half a quarter of an ounce, the next a quarter of an ounce, the next half an ounce, the next an ounce; the next two ounces, and so every one double the weight of the next lesser, and every one from an ounce upwards, are marked with numeral letters of such number, as the pieces contains Ounces, and also every Weight marked with two letters, made thus (℞) for Troy, and are or should be marked with the *Lyon* and *Leopard's Head* Crowned.

†A drachm Troy, is one penny weight and six grains; Sixteen of such drachms make an ounce Troy: Half a quarter of an Ounce is two penny weight and 12 grains.

The next Weights above them are fashioned like a Bell, and are called *Bell-Weights*, and are proportioned from one pound or 12 Ounces Troy, to 32 pounds, and sometimes higher, every

every one being double the weight of the next leſſer, as before of the ounce Weights, and are or ſhould be marked with the ſame Marks.

The Standard of theſe Weights is kept in the *Tower* of *London*, and alſo in the *Goldſmiths-Hall*, and the Officers there and none other ſhould have the Sizing or Gauging of them, but that being accounted too chargeable, the ſeveral Weight-makers in and about *London* do uſually Size and Gauge theſe Weights themſelves, according to the aforeſaid Standard, and do ſet Marks on them ſomething reſembling the right; but by what authority they ſo do, I leave to the conſideration of thoſe immediately concerned therein.

There are other ſorts of Weights, by ſome uſed amongſt us, called by the name of *Venice*-Weights, and are made in neſts of the ſame faſhion, as the neſts of *Troy* Ounces are, and every one double the Weight of the next

leſſer,

lesser, but very much differing from
the Troy Weights thus, (*viz.*) as the
Troy Ounce contains Twenty penny
weights, so the *Venice* Ounce contains but Thirteen penny weight and
a half : But there being no Law
for these *Venice* Weights amongst us,
only the Sellers of Gold and Silver
Lace (but without any warrant or
authority so to do) do too often for
their private lucre, use the same.
But what I have before mentioned of
them , is sufficient to prevent their
being used instead of the *Troy*
weights.

There are also other sorts of weights
(by our Law) in use amongst us, called by the name of *Averdupois* (the
lesser sort of them) are made of Brass,
and shaped round and flat, and every
one double the weight of the next
lesser, and are or ought to be Sized
and Marked at *Guild-Hall*, *London*,
(where the Standard of those Weights
is kept, with several Marks, (*viz.*) the
City Arms in a Shield, the Dagger,
the

the Letter A for *Averdupois*, a Flower-
de-luce, and the Effigies of a Vessel or
Ewer. These Weights differ from
the Troy Weights thus, *(viz.)* that
as the Troy Ounce contains 20 penny
weight, so the Ounce *Averdupois* contains but 18 penny weight; and as
the pound Troy contains Twelve Ounces, so the pound *Averdupois* contains Sixteen Ounces; But these not
being for the weighing Gold and Silver, what I have mentioned of them,
is sufficient to distinguish them from
the Troy Weights.

H 3   Postscript.

Postscript to the Reader.

THat my good intent for preventing fraud not onely in cases where Massy Gold and Silver are used, but also in other Works made or pretended to be made thereof, may take the better effect, I shall give you another Statute still in force, made as well to suppress and prevent the gilding and silvering of Copper and Brass Works, and the deceit

ceit therein used, as the wast-
ing the Gold and Silver of
this Nation, (viz.)

Stat. 5. *Hen.* 4. 13.

ITem. Whereas many fraudulent
Artificers imagining to deceive the
common people, do daily make
Locks, Rings, Beads, Candlesticks,
Harness for Girdles, Hilts, Challices and Sword-Pummels, Powder-
Boxes, and Covers for Cups, of
Copper, and of Lattin, and the same
over Guilt and Silver, like to Gold or
Silver; And the same sell and put in
gage to many Men not having full
knowledg thereof for whole Gold and
whole Silver, to the great deceit, loss
and hinderance of the common people,
And the wasting of the Gold and Silver, It is Ordained and Established,
That no Artificer nor other man whatsoever he be from henceforth shall gild
nor silver any such Locks, Rings,
Beads, Candlesticks, Harness for
P 4 Girdles,

Girdles, Challices, Hilts, nor Pummels for Swords, Powder-Bores,
nor Covers for Cups made of Copper or Lattin upon pain to forfeit to
the King One hundred shillings, at
every time that he shall be found guil-
ty, and to make satisfaction to the
party grieved for his damages; But
that (Chalices alwayes excepted) the
said Artificers may work or cause to be
wrought Ornaments for the Church of
Copper and Lattin, And the same
Gilt or Silver, so that alwayes in the
foot, or in some other part of every such
Ornament so to be made, the Copper
and the Lattin shall be plain, to the intent that a man may see whereof the
thing is made, for to eschew the deceit
aforesaid.

By which Statute the Guilding or
Silvering any Locks, Rings, Beads,
Candlesticks, Harness (that is, the
Buckles) for Girdles, Chalices, Hilts,
Pummels for Swords, Pouder-Boxes
and Covers for Cups made of Copper
or Lattin (to wit, Brass) is positively
forbid, upon the penalty of 5 l. for
                                    every

every offence. And one reason thereof appears to be this, That such working of Copper or Lattin, and Cover-
ing the same with Gold or Silver, is
not only a great wasting of the Gold
and Silver, but the occasion of a great
fraud, by making such Wares to be
in appearance of the value of whole
Gold and Silver, when indeed they
are in the substance thereof but dross
and Counterfeit. And no doubt but
some persons at this day (as well as
when that Statute was made) are or
may be deceived by such false Counterfeit Works, and perhaps (when it
is too late) experience the old Pro-
verb to be true, *That all is not Gold that
glisters.*

And it is not to be doubted, but
that the Makers of that Statute were
well informed that Copper or Brass
may be wrought into very many sorts
of Work wherein Gold or Silver may
be wrought. And because much deceit was then used in the working of
Buckles, and in Hilts and Pummels for
Swords, *&c.* as well as of Later times,
                                    such

such severe penalties were laid on the Workers of such false Wares, purposely to suppress all Counterfeit and deceitful Work;

Such Works being not only a wrong to the Wearer by doing little or no Service, (by reason of its brittleness, and its Kankering & stinking quality, and soon losing its Gold or Silver Colour) but also to the workers af whole Gold and whole Silver Works, who are much hindered and damnified in their Trades, by reason such false Works are made to resemble their true or right Works, and sold and worn instead thereof.

Therefore it may be reasonably conceived, that the Workers of whole Gold and whole Silver Works are intended by the said Statute, parties grieved as well as the Wearer, and shall recover satisfaction from the Offenders.

And I am well assured, that the prosecution of the offenders against

this

this Statute (which may be done by Pill, Plaint or Information) belongeth to the Wardens and Company of *Goldsmiths, London*, as well as the private agrieved party.

And the Makers and Workers of such Counterfeit Works are subjected to the Wardens and Companies power to Correct and Punish in as full and ample manner as such Workers who cover Tin with Silver mentioned in their Charter.

I Will further add something that may be useful to them that know it not, to prevent their being deceived with the Counterfeit Coyn, that (notwithstanding the great Care used by our Governors to prevent it) is continually made and vended amongst us. Wherefore they are to Observe,

First, That there is one sort of unlawful Money, that is made so, (of the lawful

lawful Coyn) by Clipping or Filing, or both, or otherwise lightening or impairing the same.

Secondly, There is another sort of unlawful Counterfeit Money, made with the mixture of a little Silver and other baser Metal, which by artificial working, boyling and finishing, will be made much like in Countenance, and near as passable as good Money.

Thirdly, There is another sort of unlawful Money, made of solid Copper or Brass, and covered or cased over (on the flat parts as thick as paper, and on the edges near as thick as a six pence) with good Silver, and is commonly as passable as the other.

Fourthly, There is another sort of Counterfeit Money made only of fine hardened Tin, which comes near (at first making) to the colour of good Silver Money.

For the knowing of these (one from another and all) from the Lawful Coyn, Note further,

Of the First, The good Silver Coyn that is clipt, filed or lessened, although thereby made unlawful, yet according as it is more or less by these ways abused, the people do receive or refuse it, as they can agree.

Of the Second, This sort, although when artificially finisht, it much resembles the proper Silver Coyn; yet it cannot be so cunningly done, but it may be discovered by its different aspect from the good; if that give cause of suspition, rub either the edge or flat part of it upon a dry board that hath gravelly or gritty matter on it, as the step of a stair, or such like; or else when the surface or out-part of the edge is a little rubb'd off, rub it on a good clean Touch-stone, as afore is directed; And if it be false, you will thereby discover it: Or else take a Goldsmith's Graver made sharp, and enter it in some part of the flat of the

Money

Money as if you begun to engrave it, and in that hole or entrance (by viewing it in the light) if bad, it will appear in its yellowish colour.

Of the Third. This fort is alwayes different from the good Silver Money thus, (to wit) as the good Silver Money hath frequently (I mean the old Money) fmall cracks on the edges thereof, which is made by the forging it; fo this fort of Counterfeit Money is generally fmooth on the edges without fuch Cracks; and as the good Money will (being let fall or thrown flat on a folid Board) ring fhrillifh; fo this fort of bad Money, by reafon its Cafe of Silver cannot be fo united to its body of Brafs or Copper, but it will (if let fall or thrown as aforefaid) found like Lead, where thefe figns give caufe of fufpition, the ufe of a Graver, as aforefaid, will plainly difcover it.

Of the Fourth. This fort is eafily difcovered thus, (that is to fay) it cannot be avoided but its countenance will look much duller or darker than
the

the good Money ; and if you bite it, you may make greater impreſſions thereon with your teeth than on the good Coyn, for it is ſofter, and much eaſier to be bent (either between the teeth, or in ſome Chink ¦or Joynt of boards) then good Money.

There may be other ſorts of Counterfeit Silver Money, and other ways to diſcover the ſame. But to enlarge this Diſcourſe on the Critick-Niceties of what is, or may be obſerveable therein, and the Laws in force concerning the ſame, would not onely ſwell this, but be matter for another Volume : Therefore I will only add, That Ingenuous perſons , being well acquainted with the Particulars of the foregoing Treatiſe, and theſe plain Rules here laid down, And by their care to obſerve nicely, the *proportion, ſtamp, countenance, and wearing,* of the good Money , will eaſily diſcern its difference from the Counterfeit. And as the *Counterfeiting, Clipping, Rounding, Waſhing, Filing, Impairing, Diminiſhing, Falſifying, Scaling or Lightening,*

ing, (for wicked lucre fake) the proper Money or Coyns of this Realm, is High Treason, by these Statutes, (to wit) 25 *Ed.* 3. 2. and 5 *El.* 11. and 18 *El.* 1. And the actors of these Crimes, their Counsellers, Consenters, or Ayders therein, to be punished accordingly: So the counterfeiting the Sterling or Standard Gold or Silver of this Realm, in any Works or Wares whatsoever, both the working, selling, exposing to sale, exchanging or bartering thereof, is severely punishable, as is afore made manifest: And it may be said of the unlawful Money, as of the unlawful Gold and Silver Works, That if the People would (as they may) be so Ingenuous to know and refuse them, the maker of either would soon desist from such unlawful Imployments.

Here

Here followeth a *Catalogue*
of the Names of the feveral Forraign Silver Coyns that are brought
into this Kingdom, as Bulloin, with
the particular weight of each Coyn,
And their particular Allay and Value, according to the aforefaid Standard of 11 oz 2 *dwt.* accounting
the fame at 5 *s.* the Ounce Troy.

Being very ufeful for all Workers in
Silver, or Traders in the faid Mo-
neys, for their ready knowing the
worth of every of them, without
being at the trouble or charge of
making Affays thereof.

I *Holland*

		dwt.	gr.
Holland *Dollar*	oo	18	5
Lyon *Dollar*	oo	17	18½
Duckstoon of Flanders	o1	oo	22
Rix-Dollar of the Empire	oo	18	15
Mexico *Real*	oo	17	12
Sevil *Real*	oo	17	12
Old Cardecu	oo	o6	3½
French *Lewis*	oo	17	11
Double Milrez of Portugal	oo	14	4
Single Milrez of Portugal	oo	o7	2
St. Mark of Venice	oo	10	4
Double Dutch Styver	oo	o1	o
Cross Dollar	oo	18	oo
Zealand Dollar	oo	13	oo
Old Philip Dollar	o1	2	o
Ferdjnando *Dollar* 1623	oo	18	6
Prince of Orange Dollar 1624	oo	18	6
Leopoldus Dollar 1624	oo	18	2
Rodolphus *Dollar* 1607	oo	18	7
Maximilian *Dollar* 1616	oo	18	2
Danish Dollar 1620	oo	13	o
Portugal Teston	oo	o5	
The Quarter of a new French *Lewis*	oo	o4	9

Is in weight

	dwt.			s.	d.
is worse	00	10	00	4	4
is worse	2	3	00	3	$4\frac{1}{2}$
is better	00	04	$\frac{1}{2}$	5	4
is worse	00	07	$\frac{1}{3}$	4	$5\frac{3}{4}$
is standard	—	—	—	4	$4\frac{1}{2}$
is better	00	01	00	4	$4\frac{1}{4}$
is worse	00	01	00	1	$6\frac{1}{4}$
is worse	00	00	$\frac{1}{2}$	4	$4\frac{1}{4}$
is worse	00	01	$\frac{1}{2}$	3	$6\frac{1}{4}$
is worse	00	01	00	1	9
is worse	00	01	$\frac{1}{2}$	2	6
is worse	04	06	00	0	$1\frac{3}{4}$
is worse	00	12	00	4	$2\frac{1}{2}$
is worse	02	00	00	2	3
is worse	01	00	00	5	0
is worse	00	12	$\frac{1}{3}$	4	3
is worse	00	10	$\frac{1}{3}$	4	$3\frac{3}{4}$
is worse	00	9	$\frac{1}{2}$	4	$3\frac{1}{4}$
is worse	00	10	00	4	4
is worse	00	4	$\frac{1}{2}$	4	$4\frac{3}{4}$
is worse	00	13	00	2	$11\frac{1}{4}$
is worse	00	01	00	1	$2\frac{3}{4}$
is worse	00	00	$\frac{1}{2}$	1	1

Is valued at

FINIS.